Governing Youth Politics in the Age of Surveillance may be one of the best books available documenting and analysing how the war on youth has become an international issue. Global in its reach, intellectually brave, and theoretically unsettling, this is a book that everyone should read if they are concerned about what is happening to youth in a world in which authoritarianism is on the rise.

–Henry Giroux, Professor for Scholarship in the Public Interest,
McMaster University, Canada

This timely edition of critical essays analyses political (re)action and resistance from young people to the growing inequality and injustices of neoliberal societies, and the state's intensified effort at suppressing them through increased surveillance.

–Paddy Rawlinson, Associate Professor of International Criminology,
Western Sydney University, Australia

Mai '68, Puerta del Sol, Chilean, Maple and Arab Springs ... Youth politics has always been a driving force of democratization and emancipation. The indispensable Grasso and Bessant's *Governing Youth Politics in the Age of Surveillance* sheds light on evolving strategies from governments that sacralise youth but penalize young people. Truly essential!

–Marie-Christine Doran, Associate Professor, School of Political Studies,
University of Ottawa, Canada

Commitments in many countries to strengthening the political participation of young people, to combat the so-called democratic deficit, have been paralleled by the growth of ubiquitous surveillance of that behaviour. Grasso and Bessant's important text documents these developments and demonstrates the importance of remaining on our guard, distinguishing the warm rhetoric of youth participation from the cold realities of state – and non-state – intervention and control.

–Howard Williamson CBE CVO, Professor of European Youth Policy,
University of South Wales, UK

This book aims to debunk and add alternative and progressive perspectives, to the response of young people as they broaden and deepen their political and civic engagement under the stronghold of restricted government policies of surveillance and criminalization. It is a narrative that must be told and there is no one better to tell it.

–Dana Fusco, Professor of Teaching Education,
City University of New York, USA

This book provides important insights into contemporary youth oriented groups involved in various kinds of political dissent. A particular strength is the rich breadth and depth of the cases which collectively provide timely insights into how campaigners are mobilizing across the world – and how state institutions and other agents of power are responding.

–Dominic Wring, Professor of Political Communication,
Loughborough University, UK

Governing Youth Politics in the Age of Surveillance

Drawing on case studies from around the world, contributors to this groundbreaking book explore a major contemporary paradox: on the one hand, young people today are at the forefront of political campaigns promoting social rights and ethical ideas that challenge authoritarian orders and elite privileges. On the other hand, too many governments, some claiming to be committed to liberal-democratic values, social inclusion and youth participation are engaged in repressing political activities that contest the *status quo*.

Contributors to this book explore how, especially since 9/11, governments, state agencies and other traditional power holders around the globe have reacted to political dissent authored by young people. While the 'need' to enhance 'youth political participation' is promoted, the cases in this book document how states are using everything from surveillance, summary offences, expulsion from universities, 'gag laws' and 'antiterrorism' legislation, and even imprisonment to repress certain forms of young people's political activism. These responses diminish the public sphere and create civic spaces hostile to political participation by any citizen.

This book forms part of *The Criminalization of Political Dissent* series. It documents and interprets the many ways contemporary governments and agencies now routinely use various techniques to repress and criminalise political dissent.

Maria T. Grasso is Professor at the Department of Politics, University of Sheffield, United Kingdom. She is the author of *Generations, Political Participation and Social Change in Western Europe* (2016) and co-editor of *Austerity and Protest: Popular Contention in Times of Economic Crisis* (2015). Her research focuses on political sociology and political engagement.

Judith Bessant is Professor at RMIT University, Melbourne, Australia. She is widely published with books in policy, sociology, politics, youth studies, media studies and history, and has worked as an advisor for governments and non-government organisations. In 2017, she became a Member of the Order of Australia for her significant service to education as a social scientist, advocate and academic specialising in youth studies research.

The Criminalization of Political Dissent

Series editor: Professor Judith Bessant
RMIT University, Melbourne, Australia

This challenging new book series explores the way governments since 9/11 from across the political spectrum intensified their efforts to criminalize both traditional and new forms of digital political dissent. The series will feature major contributions from the social sciences, law and legal studies, media studies and philosophy to document what happens when governments choose to regard activists campaigning for increased government transparency and accountability, environmental sustainability, social justice, human rights and pro-democracy as engaging in illegal activities.

The book series explores the legal, political and ethical implications when governments engage habitually in mass electronic and digital surveillance, outlaw freedom of movement real and virtual public assembly and prosecute digital activists. The books in this series are a 'must read' for anyone interested in the future of democracy.

Series titles include:

Forthcoming:

Governing Youth Politics in the Age of Surveillance
Edited by Maria T. Grasso and Judith Bessant

Shooting the Messenger: Criminalising Journalism
Authored by Andrew Fowler

Governing Youth Politics in the Age of Surveillance

Edited by
Maria T. Grasso and
Judith Bessant

Routledge
Taylor & Francis Group

LONDON AND NEW YORK

First published 2018
by Routledge
2 Park Square, Milton Park, Abingdon, Oxon OX14 4RN

and by Routledge
711 Third Avenue, New York, NY 10017

Routledge is an imprint of the Taylor & Francis Group, an informa business

© 2018 selection and editorial matter, Maria T. Grasso and
Judith Bessant; individual chapters, the contributors

British Library Cataloguing in Publication Data
A catalogue record for this book is available from the British Library

Library of Congress Cataloging in Publication Data
A catalog record for this book has been requested

ISBN: 978-1-138-63012-3 (hbk)
ISBN: 978-1-315-20974-6 (ebk)

Typeset in Goudy
by Wearset Ltd, Boldon, Tyne and Wear

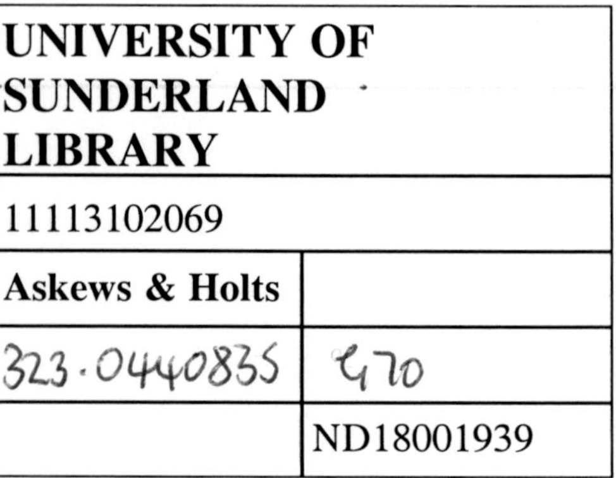

Printed and bound in Great Britain by
TJ International Ltd, Padstow, Cornwall

Contents

Figures

Contributors

Fathima Azmiya Badurdeen is a PhD candidate at the Department of Social Sciences, Technical University of Mombasa. She has been working as a researcher in the field of terrorism and forced migration in Kenya and Sri Lanka with research fellowships at the Refugee Studies Centre, Oxford University and the Mahanirban Calcutta Research Group, India.

Judith Bessant is a professor at RMIT University, Melbourne, and Adjunct Professor in the School of Justice, QUT, Brisbane. She is widely published with books in policy, sociology, politics, youth studies, media studies and history, and she has worked as an advisor for governments and non-government organisations.

Kerman Calvo (PhD, Essex University) is a senior lecturer in sociology at the University of Salamanca (Spain). His work has been published by *Sexualities, The Journal of Public Administration Theory* and *South European Society and Politics*. His latest book is *Revolución o Reforma: La Transformación de la Identidad Política del Movimiento LGTB en España* (forthcoming, Madrid: CSIC).

Luc Chicoine is a PhD student in sociology at UQAM. His thesis focuses on the various forms of repression deployed on university campuses from a comparative perspective. He is also a member of the Canada Research Chair in Sociology of Social Conflicts and a member of the Collectif de Recherche Interdisciplinaire sur la Contestation (CRIC).

Vicki Coppock is Professor of Childhood Studies and Mental Health in the Department of Social Sciences at Edge Hill University, UK. Her research and publications focus predominantly on critical social scientific analysis of theory, state policy and professional practice in childhood and youth, with a particular emphasis on asserting a positive human rights agenda for children and young people.

Laura María Fernández de Mosteyrín is a lecturer in sociology at Universidad a Distancia de Madrid. She is a member of the Group for the Studies in Society and Politics (UCM-UNED). Her research focuses on how security shapes state–citizen relations and she teaches in the field of sociology of deviance and security policy.

Maria T. Grasso is Professor at the Department of Politics, University of Sheffield, United Kingdom. She is the author of *Generations, Political Participation and Social Change in Western Europe* (2016) and co-editor of *Austerity and Protest: Popular Contention in Times of Economic Crisis* (2015). Her research focuses on political sociology and political engagement.

Surinder Guru is Lecturer in Social Work in the Department of Social Policy and Social Work at the University of Birmingham, UK. Her primary research focus is on the intersections of race, gender and class. Current work centres on UK family processes in general, and the impact of counterterrorism polices on families of detainees held under terrorist legislation in particular.

Matt Henn is Professor within the School of Social Sciences, Nottingham Trent University. He is Research Coordinator for Politics and International Studies, and academic co-ordinator for the school's doctoral programme. His research focuses on youth and political engagement and he teaches in the areas of citizenship and political participation.

Pedro Limón López has a PhD in Political Science. He is a member of the Group for the Study of Space and Power (Universidad Complutense de Madrid). His research focuses on urban geography, social movements, gentrification, and securitisation of public space.

Jane McDonnell is Senior Lecturer in Education Studies at Manchester Metropolitan University. Her research interests lie in democratic education and the arts, and the philosophical contributions of Jacques Rancière to thinking on the intersections between art, education and politics.

Norhafiza Mohd Hed recently completed her PhD at the Department of Politics, University of Sheffield. Her research focuses on the dynamic patterns of youth political (dis)engagement, particularly in the context of Malaysia and the Southeast Asian region.

Anisa Mustafa is Research Fellow at the School of Sociology and Social Policy, University of Nottingham. She has an ESRC-funded PhD (2015) from the University of Nottingham. Her doctoral thesis examines the engagement of young British Muslims in social movements to challenge Islamophobia post 9/11.

Pramod K. Nayar teaches at the Department of English, University of Hyderabad, India. His most recent books in include *Citizenship and Identity in the Age of Surveillance*, *The Extreme in Contemporary Culture*, *Human Rights and Literature* and *The Indian Graphic Novel*. His interests span culture, colonial discourse, human rights, posthumanism and literature.

Ben Oldfield is Senior Lecturer in Psychology within the School of Social Sciences at Nottingham Trent University. His research focuses on cyberpsychology, in particular digital communications, and gaming. He also works across subject areas of research using his data-handling expertise.

Sarah Pickard is a senior lecturer at Sorbonne Nouvelle University. Her research focuses on the interaction between youth policy, youth politics and youth protest in Britain. She is the author of *Politics, Protest and Young People. Political Participation and Dissent in Britain in the 21st Century*.

Martín Portos is Postdoctoral Fellow at COSMOS, Scuola Normale Superiore (Florence). He completed a PhD in Political and Social Sciences at the European University Institute in February 2017, with a thesis focused on anti-Austerity protests in Southern Europe.

Anna Schwenck is a PhD candidate and cultural sociologist at Humboldt University, Berlin. Her PhD project explores the interplay between governmental discourse, state youth politics and micro-cultural identifications amongst conservative Russian youth activists. It illuminates the ways in which political legitimacy is generated, and the complexities of young adults' apparent conformism within an authoritarian political figuration.

Paromita Sen is a doctoral candidate at the University of Virginia, USA. Her dissertation focuses on why governments choose to support and eventually co-opt social movements that are centred on issues of gender. Other research projects include the creation of online counter publics by women, women leaders in international crises and the impact of social movements on legislative behaviour.

Tony Stanley is Chief Social Worker for Birmingham City Council, UK. He is professional lead for quality social work and improving practice to help vulnerable children and their families. He is currently researching how social workers construct 'family' and 'risk' in radicalisation cases, problematising the role of statutory services in cases of suspected 'radicalisation risk'.

Rob Watts is Professor of Social Policy at RMIT University in the School of Global Studies, Social Science and Planning. He has also written or coauthored many books including *States of Violence and the Civilising Process: on Criminology and State Crime* (2016) and (with Judith Bessant and Rys Farthing) *The Precarious Generation: A Political Economy of Young People* (2017).

Acknowledgements

We are very grateful to the authors for contributing their insightful analyses of the issues and for very patiently working with us to revise and refine their chapters for the final version. We are also grateful to the many young people whose lives and political activity form the basis of this book – and thank them for the work they are doing and their commitment to enriching the political cultures in which they live. Many thanks also go to Gerhard Boomgaarden, Alyson Claffey and Diana Ciobotea at Routledge for all their help and their patience. Finally, we would like to express our gratitude to our respective universities which employed us while we edited this book – the University of Sheffield and RMIT University – as well as our families, our good colleagues and students who have supported us through this project.

Abbreviations

ACPO	Association of Chief Police Officers
AEP	Attenuating Energy Projectiles
ASBO	Anti-Social Behaviour Order
CCHQ	Conservative Party Campaign Headquarters
CND	Campaign for Nuclear Disarmament
ECHR	European Convention on Human Rights
EMA	Education Maintenance Allowance
FITs	Forward Intelligence Teams
GLC	Greater London Council
HCJ	High Court of Justice
HMIC	Her Majesty's Inspectorate of Constabulary
IPCC	Independent Police Complaints Commission
IRA	Irish Republican Army
JCHR	Joint Committee on Human Rights
LAPCC	London Assembly Police and Crime Committee
MLA	Member of the London Assembly
MP	Member of Parliament
MPA	Metropolitan Police Authority
MPS	Metropolitan Police Service
NCAFC	National Campaign against Fees and Cuts
NDNAD	National DNA Database
NUS	National Union of Students
PNC	Police National Computer
PND	Police National Database
SOCPA	Serous Organised Crime and Police Act
StWC	Stop the War Coalition
SWP	Socialist Worker Party
TSG	Territorial Support Group
UKYCC	United Kingdom Youth Climate Change coalition
UN	United Nations

Part I

Dissent and democratic practice

Governing youth politics in the age of surveillance

Judith Bessant and Maria T. Grasso

Drawing on case studies from around the world, contributors to this book explore a major contemporary paradox: on the one hand, young people today are at the forefront of political campaigns promoting social rights and ethical ideas that challenge authoritarian orders and elite privileges. On the other hand, too many governments, some claiming to be committed to liberal-democratic values, social inclusion and youth participation are engaged in repressing political activities that contest the status quo.

Contributors to this groundbreaking book explore how, especially since 9/11, governments, state agencies and other traditional power holders around the globe have reacted to political dissent authored by young people. While the 'need' to enhance 'youth political participation' is promoted, the cases in this book document how states are using everything from surveillance, summary offences, expulsion from universities, 'gag laws' and 'antiterrorism' legislation, and even imprisonment to repress certain forms of young people's political activism. These responses diminish the public sphere and create civic spaces hostile to political participation by any citizen.

This book forms part of the 'Criminalization of Political Dissent' book series. It contributes to that project by documenting and interpreting the many ways contemporary governments and agencies now routinely use various techniques from mass surveillance (much of it extra-legal), to extending the reach of criminal law and introducing tough antiterrorism laws through to violent physical repression. Contributors explore how governments across the political spectrum are 'criminalising' and proscribing traditional dissent and new kinds of political engagement.

The focus is on young people because they are playing a key role in re-generating new kinds of politics (Pickard and Bessant 2017). This, in part, is because young people have been bearing the brunt of hardship brought about by decades of neo-liberal policies in ways that older generations are not (Grasso 2018). Young people are adversely affected by increasing social inequality, high unemployment and joblessness, soaring education debt and unaffordable housing (Bessant, Farthing and Watts 2017). Young people more than most other groups are also being affected by radical changes in the labour market as human labour is displaced by digital labour, information capital and robotic technologies. All this is contributing to a growing disaffection, disillusionment and 'civil unrest' that are said to be expressed particularly by young people. This in turn elicits expressions of concern from many governments and intra-state agencies like the International Monetary Fund, the World Economic Forum and the OECD who treat young people's disaffection as evidence of a threat to 'the political consensus' (Lipton-IMF 2017; WEF 2017; OECD 2017).

As a large body of research now testifies, digital media is being used by many young people to mobilise support for a range of political agendas as they play lead roles in expanding the repertoire for political action in ways that are also transforming the political landscape. The advent of new technologies has also brought new constituencies into being, and novel opportunities to participate in ways they have not previously been able to. The new digital and social media has enabled young people who have historically been marginalised or even excluded from the public sphere to participate in increasing numbers.

The cases documented here point to important challenges to the well-established conventions about the relations between citizens and their governments. They raise questions about whether we can take for granted democratic rights like freedom of expression and movement; privacy and the rule of law principles are still important in the context of the rise of the 'security state'.

Governing young people

Young people have long been one of the most intensely regulated groups. They have long been subject to various forms of governance of different kinds in ways that would not be tolerated if they were applied to any other group (Rose 1999). Whatever else has changed since Nikolas Rose made this observation in the late twentieth century, the contemporary ways of seeing and responding to young people and particularly their politics, points to strong continuities with the persistent interest in their governance.

As Foucault argued, the capacity to govern a group relies on the ability to shape the ways those categories of people and social reality is known and talked about (1979). That is, how we categorise and determine how certain individuals and groups like 'youth', 'the unemployed', 'terrorists' and so forth are named and represented is the first step in their governance.

Concern about young people's political engagement – or 'disengagement' – is not new (for detailed discussion of the issues at hand in this respect see Grasso, 2011, 2013, 2016). Indeed, contemporary arguments about the various threats ostensibly posed by the political action of some young people bear a conspicuous resemblance to older interests in the governance of young people that have deployed a full range of the theoretical models of deviancy, to delinquency to its more recent incarnation 'youth at risk'.

When young people act politically in ways that fall outside the domain of state-expert sponsored or managed forums like 'youth round tables' or similar devices, the response of governments typically involves retreating to an older discursive tradition about the inherently troubled and troublesome nature of 'youth' and the need for their governance. Historians like Pearson document the long tradition of popular fears and the way each generation believes itself uniquely threatened by new forms of degeneration caused by the proclivities of young 'transgressives' (Pearson 1983). Moreover, the defining features of youth-adolescent in the West include being crisis-ridden, troubled, socially irresponsible and defiant. As such, 'youth' are constantly 'at risk' of being diverted off that rickety and perilous 'transitionary path' towards responsible adulthood and towards irresponsibility, and even criminality (Stanley Hall 1905). By 1942, the sociologist Talcott Parsons was defining 'youth culture' in terms of 'youthful irresponsibility' (Parsons 1942/1963). It is an account that has been reinforced by various experts and mainstream media reports that represent young people as 'trouble' because of their supposed inherent dispositions towards 'youthful disorder'. As 'folk devils in the making' or 'lords of misrule', they have long provided copy for media-sponsored 'moral panics' (Cohen 1979). By the 1990s, a new discourse of 'risk' had silently taken over the old discourse (Bessant *et al.* 2003). Yet, while there are certainly challenges to be met in being young, particularly in the contemporary context, young people are no more troubled or troublesome than any other sector of the population.

Theorising surveillance

As contributors to this book highlight, the governance of young people and their politics now implicates states, private corporations and even universities and schools. It entails a full range of regulatory and management techniques and crucially, it now includes the gathering and use of unprecedented information about the lives of citizens (Chamayou 2015). Civil society long deemed to be the space of private or free practices oriented to public opinion formation and deliberation is now being subjected to permanent and total surveillance.

It will be recalled that drawing on Bentham's idea of the Panopticon as a metaphor, Foucault argued that modern society deploys surveillance to regulate the lives of citizens and promote social order (1979). Bentham's Panopticon was a design principle that enabled the observation of inmates of a prison or some other institution while eliminating the visibility of the guards. The inmates of such an institution, unsure of whether or not they were subject to guards' gaze, were always governed and thus they acted at all times as if they are watched so as to avoid punishment. This became an internalised practice as they came to govern themselves. For Foucault, the Panopticon entails a 'set of techniques and institutions for measuring, supervising and correcting the abnormal ... which, even today, are disposed around the abnormal individual, to brand him and to alter him' (Foucault 1979: 199). In this way, a Foucauldian account of surveillance provides a useful framework for theorising the impact of perceived government surveillance on young people's political activity as well as allowing for a range of reactions to surveillance, like docility or political resistance.

Mass surveillance has now become a fundamental aspect of contemporary industrialised societies carried out by states, intelligence agencies, private corporations and even universities and schools (Marsden 2014). For Staples, surveillance involves 'keeping close watch on people' or generally a 'watchful gaze' (1997: ix). As Chamayou (2015: 78) observes, we live now under a 'constant geo-spatial overwatch' by 'the institutional eyes' of Google, or the National Security Agency that monitor our everyday lives by collecting and 'processing personal data, whether identifiable or not for the purpose of influencing or managing those whose data have been garnered' (Lyon 2001). Internet-based surveillance techniques used by governments conform in many ways to this Panopticon model as it provides highly effective disciplinary means requiring a few to watch the many while preventing the many from watching the few (Boyne 2000). Conversely, internet users can also watch their watcher. As Boyne argues, 'the machinery of surveillance is now always potentially in the service of the crowd as much as the executive' (2000: 301).

Lyon observes that, increasingly, surveillance does not involve embodied persons watching each other (2001: 2). It has come to incorporate everything from visual observation by CCTV, genetic testing, the collection of biometric data and use of facial recognition systems, electronic monitoring, (e.g. opinion polling, metadata analysis of emails and phone calls), and the collection of social statistical data to observe trends in social conduct (like crime, unemployment) and so on. Most surveillance operates abstractly and at a distance: it can be characterised as a form of remote and impersonal watching. In this way, it separates watching from witnessing. And increasingly, the watching, analysing and interpreting is automated and reliant on algorithms. In these ways, surveillance has become both ubiquitous and normal.

Traditionally, when political science engaged with political surveillance, it tended to be framed as a defensible response to small, well-defined groups of dissidents or criminals (Cunningham 2004). This naivety became less feasible after 9/11 when states began to mandate mass surveillance by agencies like the National Security Agency (Scheuerman 2016). In the USA, the passage of the USAPATRIOT Act (Uniting and Strengthening America by Providing Appropriate Tools Required to Intercept and Obstruct Terrorism) removed any barriers that prohibited the government from intercepting electronic mail, net activity or mobile phone use (Berkowitz 2002; Bauman *et al.* 2014). It is now understood that governments in Europe, Australia and the USA regularly subject their own citizens and those of other nation-states to total surveillance of their electronic communications (Gardner 2016; Stoycheff 2016). Snowden's revelations highlighted how the National Security Agency, through its PRISM project, has engaged in mass online surveillance programmes since 2004 in America and globally. PRISM allows US intelligence agencies to direct albeit 'backdoor' access to companies like Google, Microsoft, Apple, Facebook, Yahoo, YouTube, Skype, and others, introducing the potential for surveillance of Americans without warrants (Greenwald and MacAskill 2013).

Surveillance has also become a normal feature of workplaces, schools, homes and communities. The once private space of the home has been well and truly entered by home-based media as we create possibilities for fully integrated surveillance systems that track our consumer and other activities. In the home, keyboard strokes are harvested, media preferences are collected, individual interactions with information technologies are now stored, tracked and logged by platforms like Facebook and Google. Meanwhile, 'data miners' monitor the 'mood' of the Internet via real-time analysis of online conversations, blog-posts, Tweets, Facebook posts and other activities.

The rise of neo-liberal economic regimes also coincided with more punitive approaches to crime and welfare. Public spaces like our streets and city squares are covered by networks of CCTV cameras and other governance technologies. Neoliberal governments also initiated the harsh and often repressive treatment of welfare claimants in a 'new paternalism' reliant on measures like 'activity tests' designed to target areas of substantial socio-economic deprivation, underpinned by a costly bureaucratic surveillance, monitoring and infrastructure of sanctions (Bessant *et al.* 2017; see also Dunn *et al.* 2014, Grasso *et al.* 2017).

The availability of new technologies that automate the collection and analysis of material in the context of the 'security state' has seen an expansion of the concept of 'security threats' to a situation where everyone becomes a potential threat.

Finally, we asked whether what is happening presents an opportunity for reflexivity by those claiming disengagement from electoral politics is evidence of young people's apathy, justifying measures to govern by mandating 'correct behaviour'.

What is and what ought to be the role of the social sciences in the debates about the 'civic crisis' (Hay 2007)? Is the modernist self-portrait of the social sciences as handmaiden to the state in helping to manage or control recurring social order problems and crises through 'scientific' research still relevant or warranted? This raises ethical-political questions about the social sciences and the 'human interests' that inform them. Are they informed by an interest in control and domination or by a preference for promoting freedom and political autonomy (Habermas 1971)?

The social sciences have become central to the normalisation of surveillance technologies now employed to govern the politics and, more generally, the lives of young people often understood as a high-risk demographic. It is an observation that invites some questions about the politics of social scientific research and intervention of professionals in the lives of young people using categories like 'youth at risk', 'problem youth' and 'social capital' to warrant actions designed to mitigate the risk.

Book structure

The chapters in this book identify how many young people are engaging in political activism as well as the responses to this by the state and other key social institutions.

In the next chapter, 'Theorising student protest, liberalism and the problem of legitimacy', Rob Watts (Royal Melbourne Institute of Technology University) helps set the scene for many of the subsequent chapters when he suggests that the problem of legitimacy is central to, though not always fully explicated in, the relationship of young people's dissent to the attempts to regulate or repress it. He draws on three case studies of student protests in Malaysia, Sussex, England and Québec, Canada between 2012 and 2014 to highlight some important issues about the commitment on the part of liberal democracies to liberal principles like freedom of expression or the right to dissent. In each case, student activism was subject to repression by university authorities and even criminalisation by states. Watts argues that liberalism, far from being preoccupied with liberty and rights, has a long-term commitment to security that always trumps freedom. The tendency of liberal-democratic states and authoritarian states to prioritise security, leads them to regard young political actors like those engaging in 'human rights activism' pro-democracy, 'anti-Austerity', anti-privatisation and freedom of speech campaigns, as antagonistic to the 'national interest' or 'public safety'. In this way, what starts as stigmatising young activists as a threat to 'national security' or 'public order', frequently ends up with activists facing criminal charges.

In the third chapter, 'Panic works: the "gag law" and the unruly youth in Spain', Kerman Calvo (University of Salamanca) and Martín Portos (Scuola Normale Superiore, Florence) identify the so-called Ley Mordaza [Gag Law] as the centrepiece of a new regime of governance of young people's politics in Spain. Their account highlights how Spanish young people played a key role in campaigns under the shadow of the 2008 economic recession, as they protested against cuts to welfare education, health programmes and other 'neo-liberal inspired' policies. They describe how conservative forces in Spain depict politically active young people as 'unruly' and threatening to the Spanish state. They document how events were framed in ways that generated popular panic and fear about 'juvenile crime' to justify new government-sponsored repression and surveillance that limited various civic rights, including the right to protest. Calvo and Portos report how the Spanish government drew on time-honoured fears of 'tumultuous youngsters' to represent young political actors engaged in legitimate political action requiring close management. Finally, they document campaigns against those 'Gag Laws' and explain why government actions failed.

Chapter 4, 'Controlling dissent through security in contemporary Spain', also looks at Spain, but from a different perspective. Authors Laura María Fernández de Mosteyrín (Universidad a Distancia de Madrid) and Pedro Limón López (Universidad Complutense de Madrid) document how Spanish governments responded to young people's political dissent in the period 1999–2016. They argue a key feature of Spanish protest actions since the 2000s has been the role of young people. Collective action involving 'alter-globalisation' ['altermondialiste'] groups, pacifist groups, student or housing movements evolved and merged to become the Indignados Movement (15M) in 2011. The authors document the recent expansion of state power to control this political action that involved extending the reach of criminal law, surveillance of public spaces and other forms of repression, (e.g. through 'counter-radicalisation strategies'). This case study also shows how the convergence of counterterrorism and public order initiatives worked to expand state power to regulate social life. The authors demonstrate how new Spanish 'Gag Laws' target those labelled 'youth radicals'. In enacting such legislation, the state developed new institutional agencies designed to regulate and manage 'conflict' while curtailing citizens' rights in the name of security and community safety.

Luc Chicoine (Université du Québec à Montréal) picks up on these themes in the next chapter, '"Proxy Repression"? The causes behind the change of protest control repertoire by the Université du Québec à Montréal during the 2015 student strike'. He documents the increasing repression of student mobilisations over the last decade, including moves by the university administration to delegitimise student claims, use electronic surveillance, and criminalise protest actions. He argues that the bid to repress the 2012 student strikes backfired in ways that saw the mobilisation of students and other sections of the population. In 2015, however, moves by the university to repress student protests against austerity were faster, and focused mostly on the campus of Université du Québec à Montréal (UQAM). They included expelling student activists and mass arrests during a class strike. Added to this was the extensive use of surveillance cameras by UQAM's prevention and security services and surveillance of students by agents of private security firms. As Luc Chicoine argues, the main rationale for these actions was to deny the legitimacy of the student voice.

The inquiry into the criminalisation of young people's political dissent is continued by Sarah Pickard (Université Sorbonne Nouvelle, Paris) in her chapter, 'Governing, monitoring and regulating youth protest in contemporary Britain'. Pickard argues that young people in Britain were already involved in political protest when the Conservative–Liberal Democrat coalition government came to power (May 2010–May 2015). She shows how youth-led protests increased in Britain as did state-sponsored governance and surveillance. It was political action that took the form of petitions, peaceful direct action, 'creative civil disobedience', street marches, peace camps, lock-ins, sit-ins, break-ins, occupations, encampments and even violence. Pickard documents the increased use of legislation designed to regulate political dissent, to monitor protest and to create a more militarised police force. Because of the strict application of existing laws and the introduction of new, tougher legislation, the police acquired greater power to surveil, gather intelligence and control crowds. For this reason, young protesters were more likely to be disciplined and receive a criminal penalty than they had been in the past.

Fathima Azmiya Badurdeen (Technical University of Mombasa), in her chapter titled 'Surveillance of young Muslims and counterterrorism in Kenya', reports on how Kenya's government responded to some Muslim young people with heavy-handed counterterrorism strategies. This occurred in the context of the Kenyan 'war on terror', where young Muslims were officially described as vulnerable to radicalisation and part of 'suspect communities'. Badurdeen's focus is on state-sponsored surveillance and its influence on Muslim young people. She explores how antiterrorism laws and policing practices have curtailed the freedom and capacity of many young Muslims to participate in civil society, while in some instances they have promoted the radicalisation of Islamic young people. She highlights the use of racial and religious profiling, antiterrorism legislation and antiterrorism measures involving surveillance of 'suspect communities' – a code for 'Muslim youth'. Recent developments like the closure of mosques and the surveillance of public and digital spaces targeting Muslims has proved to be the prelude to closing down public spaces used by young people to engage in legitimate political action.

The surveillance of young Muslims is the theme of the next chapter, 'On becoming "radicalised": pre-emptive surveillance and intervention to save the young Muslim in the UK' by Vicki Coppock (Edge Hill University, Ormskirk), Surinder Guru (University of Birmingham) and Tony Stanley (Birmingham City Council). This chapter critically analyses the impact of counterterrorism laws, policy and practice on Muslim children and young people in the UK, and specifically how British Muslims have become central to official discourses about 'vulnerability to radicalisation'. The authors critically assess the assumptions and arguments informing this vulnerability to radicalisation discourse that gives legitimacy to preventive justice measures, including surveillance and interventions in the lives of young British Muslims. As they argue, while these practices are ostensibly designed to be in their best interests, they actually pathologise and delegitimise youthful political dissent. The authors show how official accounts of young people's 'vulnerability to radicalisation' are shaped by particular ideas about the ways 'radicalisation' takes place, alongside specific constructions of the 'Islamic' 'child'. This, they argue, is a process reliant on essentialised ethnic and racial identities married to equally problematic constructions of 'childhood' and children's 'mental health and well-being'.

Anisa Mustafa (University of Nottingham) focuses on the politics of young British Muslims. Her chapter, titled 'Active citizenship and governmentality: the politics and resistance of young Muslims in the security state', Mustafa argues that in the context of intense surveillance, the citizenship of young Muslims has become contingent on their willingness to demonstrate loyalty to Britain. Based on ethnographic research in the British Midlands, she enquires into their resistance to the increasingly strident and exclusionary concept of 'active citizenship' promoted by British governments. Viewed through the lens of Foucault's concept of governmentality, 'active citizenship' is best seen as a technology of power that dampens dissent amongst citizens. Anisa Mustafa argues that young Muslims engage in activities in ways that share the repertoire of action, transformative objectives and imaginaries with social movements. She identifies two models of active citizenship: one marked by 'explicit resistance' and the other, 'implicit resistance'. Both reveal the contingencies and precariousness of Muslim citizenship in contemporary Britain under regimes of governance marked by growing securitisation and neo-liberalisation of society.

Jane McDonnell (Manchester Metropolitan University) asks, 'What future for young people's political activism?' Her chapter centres on the role of art and aesthetics in young people's political protest. McDonnell addresses the paradox that liberal-democratic states, ostensibly committed to democratic values, repress various forms of political action including, in this case, the use of artistic strategies in young people's protest. As she argues, this repression occurs not only through overt measures of criminalisation and control, but also through the more diffuse workings of power within neo-liberal states. This, in turn, highlights a further paradox: that the systems and institutions through which such power operates are also the very sites in which political dissent can have most effect. Drawing on the work of Jacques Rancière, Jane McDonnell offers an optimistic understanding of this paradox. In particular, she illustrates how young people's political dissent can turn surveillance operations on themselves. This insight has significant implications for art, aesthetics and activism.

The following chapter, 'Effects of the regime in Malaysia on youth political participation' by Norhafiza Mohd Hed (University of Sheffield), examines efforts by the Malaysian government to repress and criminalise political dissent. She asks what effect the Malaysian government's efforts to repress democratic dissent have had on young Malaysians' willingness to engage politically. Based primarily on interviews with young Malaysians, Norhafiza Mohd Hed reports how state-sponsored repression, particularly the ongoing use of criminal laws and violence, has created a 'culture of fear' and is seen as deterrent for many young people. This chapter reports how interviewees felt inhibited and were fearful in ways that made them disengage from formal and informal political participation, including protest action. The Malaysian case raises questions about the political role of young people wanting to pursue democratisation while living with a government committed to repressing such a project. In this regard, fear that their political actions will also be criminal actions has seen many young Malaysians see political engagement as too high-risk. In this way, the findings are broadly consistent with existing literature which argues that state repression decreases citizens' interest in political participation.

In her chapter, 'Russian politics of radicalisation and surveillance', Anna Schwenck (Humboldt-Universität, Berlin) shows how, in Russia, policies relating to young people and anti-extremism have expanded considerably in recent years. She interprets this development as a radicalisation of the Russian state – its extension of surveillance and control practices. This radicalisation is a response to national threat frames (i.e. schemata of interpretation which make the national collectivity appear endangered). The analysis of the 'Orange' and 'Pink' threat frames are used to show that both feed on the figure of the (sexually) immature young adult, who is vulnerable to manipulation by alien forces. Popular public narratives stabilise these threat frames and lend the two disciplining policy fields legitimacy.

The next chapter, 'Biocultural metrics and the moral policing of young people's politics in contemporary India' by Pramod K. Nayar (University of Hyderabad), examines the moral policing of young people's politics in contemporary India. He argues that a 'biocultural metric' evolved as part of the process of surveilling young peoples' bodies, that was then used to identify them as 'anti-national' or 'unIndian'. As Nayar explains, this has redefined public space, and the idea of 'the public' in India. In this way, young people 'asserting themselves' in public spaces have been monitored and judged according to particular cultural criteria. This process has undermined young people's rights and ability to use the public space. In this way, the restriction of young people's behaviours in contemporary India symbolises attempts by vigilante groups of various religious and ethnic persuasions to assert their own models of Indian citizenship. These models are monocultural responses to the forces of globalisation, hybridisation and the contingent politics of the postmodern. The guarding of cultural boundaries through surveillance in this way reconstitutes them as a problem.

In her chapter examining 'Surveillance and the student: government policing of young women's politics', Paromita Sen (University of Virginia, Charlottesville) focuses on the measures the government of India uses to police young women's politics, like the women's movement #PinjraTod which is demanding equitable living conditions for male and female college students across the country. #PinjraTod came into existence in September 2015 to ensure 'secure, affordable and not gender-discriminatory accommodation for women students across Delhi' and since then has spread across the country. While discriminatory policies have been the norm for many years, advances made since the 1990s were reversed in the aftermath of the now infamous 2012 brutal rape and murder of a college student in Delhi. That case allowed university administrations to roll back civil liberties, justified to parents under the guise of protecting their students. Female students now face a range of discriminatory rules and regulations that attempt to control and limit their access, mobility, sexuality and lived experiences, including but not limited to curfews, price discrimination, moral policing and punitive threats of university sanctions. These measures rely on traditional techniques as opposed to reliance on digital surveillance – such as curfews, moral policing and dress codes. A reliance on traditional techniques draws on the strength of familial and social networks to help police and control dissent amongst young women on college campuses and thereby cement the role of women in the private sphere.

The final chapter by Matt Henn, Ben Oldfield (Nottingham Trent University) and Judith Bessant (RMIT University, Melbourne) addresses a central issue in contemporary politics: 'Electoral engineering and surveillance: British young people and politics' and reports on how many young people in Britain are disengaging from formal electoral politics while increasing their participation in political activism. Data collected reveals that many young people feel disconnected from formal political processes. The authors consider whether electoral engineering methods – designed to persuade or compel young people to vote – can be understood as a form of political governance and surveillance. They go on to argue that while age-specific compulsory voting may increase the numbers of young voters, it may also have negative consequences. They also argue that if young people are to be encouraged to participate in formal politics, the onus is then on political elites to consider how democratic participation can be made more meaningful for young people.

Conclusion

While acknowledging that state criminalisation of activities seen to threaten certain conceptions of the social or political order or the interest of elites is hardly new, we ask how we might best understand contemporary moves to criminalise political dissent involving young people. We suggest there are four points worth considering.

First, we argue there is value in resisting claims in the commentary by many politicians, policymakers and in the research by some academic social scientists that a reluctance or refusal to vote or a disenchantment with politics by many young people is best understood as evidence of their apathy or that it is due to some kind of generational nihilism. Rather, as authors in this book and evidence documented elsewhere highlight (e.g. Pickard and Bessant 2017), large numbers of young people are deeply committed ethically and politically to the communities and world in which they live (see also Giugni and Grasso 2015a). Moreover, the anger and dismay directed at 'normal politics', at the major parties and politicians, does not just come from young people, but from people across the generations who are increasingly dismayed by the quality of the official rhetoric and the disposition to 'lead' by relying on the latest public opinion polls or tabloid media headlines.

Second, behind that growing disappointment is a recognition that decades of neo-liberal policymaking and its preoccupation with privatisation, deregulation, tax cuts for companies and high-income earners, cuts to welfare, user-pays to fund education, health and infrastructure are not working and are causing considerable social harm and that the neo-liberal policy experiment has failed to deliver sustainable deep economic growth and fair societies. It has contributed to a major 'legitimation crisis' (Habermas 1974). As Habermas argued, such crises start when the state is expected to steer the economy to meet the needs of capital accumulation, while simultaneously implementing health, welfare education and income redistribution policies to help create a more equal society. However, this is something which cannot be achieved if the requirements of capital accumulation are to be met (Plant 1983: 341).

The result is a 'crisis of democracy' now affecting the political system which can no longer be ignored, especially after the 2008 Global Recession (Giugni and Grasso 2015b; Grasso and Giugni 2016). Piketty's (2014) intervention provides empirical verification and detail of how the last decades have seen income inequality in many developed societies return to levels we have not seen since the end of the nineteenth century. The recent recession highlighted how the financialisation of many developed economies had come to rely on Ponzi schemes and speculative bubbles, themselves based on high-risk housing mortgage loans and mysterious derivatives that were bad for the real economy and capital accumulation. The response of free-market governments was to save the financial institutions that caused the crisis using public funds and debt and then to impose austerity measures on the most vulnerable citizens who had no fault in the emergence of this crisis (Giugni and Grasso 2017).

Third, this crisis of legitimacy is why dissent matters. As Jacques Rancière argues, 'the essence of politics is dissensus' (2010: 38; also, Mouffe 2005). It is an understanding of dissensus understood not just as conflict between interests, opinions, or 'values', but as a reconfiguration of the status quo, as the rearrangement of more common experiences or what is sensible and known (Rancière 2010: 69). Dissent can play a decisive role in moving us towards new democratic political and policy paradigms and more just, humane models for society.

Finally, if liberal democracies represent themselves as champions of liberties like freedom of speech, freedom of association and the right to free and contested elections, should these states and their decision makers not then welcome social movements and dissent of all kinds as a natural and critical part of citizen participation in decision-making? (Brabazon 2006: 3). Surely, this is a good question on which to end this introduction and for its readers to consider while reading the chapters that follow.

References

Bauman, Z., Bigo, D., Esteves, P., Guild, E., Jabri, V., Lyon, D. and Walker, R. (2014). After Snowden: rethinking the impact of surveillance. *International Political Sociology*, 8(2): 121–144.

Berkowitz, R. (2002). Packet sniffers and privacy: why the no-suspicion-required standard in the USA PATRIOT act is unconstitutional. *Computer Law Review and Technology Journal*, 7(3): 1–21.

Bessant, J., Farthing, R. and Watts, R. (2017). *The precarious generation: A political economy of young people*. Abingdon: Routledge.

Bessant, J., Hil, R. and Watts, R. (2003). *Discovering risk: social research and policy making*. London: Sage.

Boyne, R. (2000). Post-panopticism. *Economy and Society*, 29(2): 285–307.

Brabazon, H. (2006). Protecting whose security? Anti-terrorism legislation and the criminalization of dissent. YCIS Working Paper 43, York University, Toronto.

Chamayou, G. (2015). *A theory of the drone*. New York: The New Press.

Cohen, S. (1979). *Folk devils and moral panics*. Second Edition. Oxford: Martin Robertson.

Cunningham, D. (2004). *There's something happening here: The New Left, the Klan, and FBI counterintelligence*. Berkeley: University of California Press.

Dunn, A., Grasso, M. and Saunders, C. (2014). Unemployment and attitudes to work: asking the 'right' question. *Work, Employment, and Society*, 28(6): 904–925.

Foucault, M. (1979). *Discipline and punish: The birth of the prison*. New York; London: Vintage Free Association Books.

Gardner, L. (2016). *The war on leakers: national security and American democracy, from Eugene V. Debs to Edward Snowden*. New York: The New Press.

Giugni, M. and Grasso, M. (2015a). Environmental movements in advanced industrial democracies: heterogeneity, transformation, and institutionalization. *Annual Review of Environment and Resources*, 40: 337–361.

Giugni, M. and Grasso, M. (eds). (2015b). *Austerity and protest: popular contention in times of economic crisis*. London: Routledge.

Giugni, M. and Grasso, M. (eds). (2018). *Citizens and the crisis: perceptions, experiences, and responses to the Great Recession in Europe*. London: Palgrave Macmillan.

Grasso, M. (2011). *Political participation in Western Europe*. DPhil thesis, Nuffield College, University of Oxford.

Grasso, M. (2013). The differential impact of education on young people's political activism: comparing Italy and the United Kingdom. *Comparative Sociology*, 12(1): 1–30.

Grasso, M. (2016). *Generations, political participation and social change in Western Europe*. London: Routledge.

Grasso, M. (2018). Young people's political participation in times of crisis. In: S. Pickard and J. Bessant (eds), *Young people re-generating politics in times of crises*. London: Palgrave.

Grasso, M. and Giugni, M. (2016). Protest participation and economic crisis: The conditioning role of political opportunities. *European Journal of Political Research*, 55(4): 663–680.

Grasso, M., Farrall, S., Gray, E., Hay, C. and Jennings, W. (2017). Thatcher's Children, Blair's Babies, political socialisation and trickle-down value-change: An age, period and cohort analysis. *British Journal of Political Science* https://doi.org/10.1017/S0007123416000375.

Greenwald, G. and MacAskill, E. (2013). 'NSA prism program taps in to user data of Apple, Google and others'. *Guardian*, 7 June. Available from www.theguardian.com/world/2013/jun/06/us-tech-giants-nsa-data [accessed 5 May 2017].

Habermas, J. (1971). *Knowledge and human interests*. Boston: Beacon Press.

Habermas, J. (1974). *Legitimation crisis*. Boston: Beacon Press.

Hay, C. (2007). *Why we hate politics*. Cambridge: Polity.

Lipton, D. and IMF (2017). Stronger cooperation to promote inclusive growth. IMF. Available from www.youtube.com/watch?v=r22CMhMGOGE [accessed 15 September 2017].

Lyon, D. (2001). *Surveillance society: monitoring everyday life*. London: McGraw Hill.

Marsden, C. (2014). Hyper-power and private monopoly: The unholy marriage of (neo) corporatism and the imperial surveillance state. *Critical Studies in Media Communication*, 31(2): 100–108.

Mouffe, C. (2005). *On the political*. London: Routledge.

OECD (2017). Key issues for digital transformation in the G20. Report prepared for a joint German Residency OECD Conference, Berlin, January. Available from www.oecd.org/g20/key-issues-for-digital-transformation-in-the-g20.pdf [accessed 15 September 2017].

Parsons, T. (1942/1963). Age and sex in the social structure of the United States. In: R. Grinder (ed.), *Studies in adolescence*. London: Macmillan, 28–49.

Pearson, G. (1983). *Hooligan: A history of respectable fears*. London: Routledge.

Pickard, S. and Bessant, J. (eds). (2017). *Young people re-generating politics in times of crises*. London: Palgrave Macmillan.

Piketty, T. (2014). *Capital in the twenty-first century*. Cambridge: Harvard University Press.

Plant, R. (1983). Jurgen Habermas and the idea of legitimation crisis. *European Journal of Political Research*, 10(4): 341–352. Available from www.caledonianblogs.net/making_the_case/files/2010/04/fulltext.pdf [accessed 28 November 2017].

Rancière, J. (2010). *Dissensus: on politics and aesthetics*. New York: Continuum.

Rose, N. (1999). *Governing the Soul: the shaping of the private self* (Second Edition). Abingdon: Routledge.

Scheuerman, W. (2016). Digital disobedience and the law. *New Political Science*, 38(3): 299–314.

Stanley Hall, G. (1905). *Adolescence: its psychology and its relations to physiology, anthropology, sociology, sex, crime, religion and education*. London: Appleton.

Staples, W. (2000). *Everyday surveillance: vigilance and visibility in postmodern life*. New York: Rowman & Littlefield.

Stoycheff, E. (2016). Under surveillance: examining Facebook's spiral of silence effects in the wake of NSA internet monitoring. *Journalism & Mass Communication Quarterly*, 93(2): 296–311.

World Economic Forum (2017). The Global Risks Report 2017. Available from www3.weforum.org/docs/GRR17_Report_web.pdf [accessed 15 September 2017].

Chapter 2

Theorising student protest, liberalism and the problem of legitimacy

Rob Watts

Introduction

On 27 August 2016, a coalition of student groups, Kesatuan Mahasiswa Malaysia, staged a rally in Kuala Lumpur called the 'TangkapMO1' or the 'Catch the Malaysian Official 1'. This rally, as every Malaysian knew, was about government corruption. The prevailing climate of fear created by state repression meant that Malaysian news reporters were not prepared to name 'Malaysian Official 1' as Prime Minister Najib Razak (Kumar and Palansamy 2016). Razak is still facing international investigations in late 2017 over how USD 700 million (AUD 975 million) in a state sovereign wealth fund made its way into his bank account (Mahriti 2017). Since this fact could not be openly acknowledged, public commentators referred to 'Malaysian Official 1'. The rally was run as planned in spite of police warnings that the protest was illegal. Nearly 1,000 people rallied, mostly students, chanting 'Tangkap MO1' ['Catch the Malaysian Official 1'] and 'Hidup mahasiswa' ['Long live the students']. The rally passed off peacefully on this occasion.

Other Malaysian student protesters fared less well. In October 2016, four University of Malaysia students, including Anis Syafiqah Mohd Yusof, the chairperson of Kesatuan Mahasiswa Malaysia, were summoned to a disciplinary hearing held in December 2016. The university disciplinary panel found them guilty of 'acting in a manner detrimental to both the university's good name and public order'. As human rights activists noted, this decision breeched Article 10(a) of Malaysia's Federal Constitution providing for freedom of speech and expression and Article 10(b) which protects the freedom to assemble peacefully. Three were subsequently suspended for a semester and fined. All appealed the judgement; as of mid-2017, the results have not been determined.

This chapter explores this and a number of other cases in which university student protesters provoked political and legal attempts by university authorities and governments to discipline or criminalise political dissent. It might be thought that what happened in Malyasia was predictable, given that Malaysian governments have long relied on a discourse about 'Asian values' to distance themselves from liberal-democratic traditions (Loh Kok Wah and Boo Tiek

2013). Yet, as this chapter suggests, cases like the Sussex University Five (2012–2015) in the United Kingdom and the so-called 'Maple Spring' in Québec, Canada (2012), saw university managements and ostensibly liberal-democratic governments criminalise dissent and take deeply repressive measures to stamp out student dissent. These cases and the responses of university administrations and governments raise large issues about the state of democratic rights, including the right to dissent, and the legitimacy of state efforts to criminalise and repress dissent. (This chapter does not focus on surveillance so much as on criminalisation and other techniques of repression.)

This chapter argues that state-sponsored repression of dissent highlights major conceptual and ethical and political problems about modern political dissent and its place in ostensibly liberal political orders. That liberal states have criminalised dissent points to one major question: on what grounds do states legitimise their use of force? As writers since Pascal (1670: 94) through to more recent figures like Dyzenhaus (1997), Schmitt (2003) and Loughlin (2015) suggest, there is no good answer to this, especially when liberals shroud their answers in the claim that they uphold 'the rule of law' (also Dyzenhaus and Poole 2015). As Tuck (2016) argues, part of the difficulty rests on an inability to distinguish between 'sovereignty' (or 'legitimacy') and 'government'. As Koskenniemi (2012) and Dyzenhaus and Poole (2015) emphasise, this relies on the long-standing inability of the Anglo-American tradition of legal positivism to treat the law as political.

A lot of attention has rightly been paid to the politico-legal problems set loose since 9/11 as states engaged in a 'war on terror' involving heightened security and 'counterterrorism' measures in spite of the absence of evidence about the actual scale of the threat represented by terrorism (Mueller and Stewart 2016). Surprisingly little, if any, attention has been given to the issues attendant on the criminalisation of dissent that this has promoted.

This relative silence is odd, given that the last decade has been marked by significant expression of dissent around the globe. The Great Recession of 2008 prompted waves of dissent including the Arab Spring in North Africa, the anti-Austerity campaigns in Spain, Greece, Portugal, Italy and Britain, as well as the UK riots of 2011, movements like Occupy Wall Street and Black Lives Matter in America, the 'Maple Spring' in Canada or the 'Chilean Spring'. More recently, dissent has been registered in other ways, including the mobilisation of anti-EU and anti-immigration sentiment in electoral processes that led to the rise of populist parties in Italy, Spain and Greece as well as the Brexit vote in Britain in May 2016 and the presidential election of Trump in November 2016.

It is odd that the way liberal states especially have criminalised many forms of dissent has not been the subject of sustained critical attention. As writers like Brabazon ask, if liberal democracies around the world pride themselves on principles like freedom of association and democratic rights such as the right to free speech and dissent, 'Why would these states and their decision makers not welcome social movements and dissent of all kinds as a natural part of the rich fabric of citizen participation in decision-making?' (2006: 3).

As is clear in the three cases examined here, governments have attempted with varying degrees of success to silence, repress and regulate dissent by making public gatherings or free speech illegal, using police violence to break up street marches or occupations, mass arrests and criminalising digital political activism. The introduction of coercive preventive justice measures and suspension of rights like *habeas corpus* has been justified by claims the 'West' is now engaged in a 'war on terror' or, as in Malaysia, with claims that stability and social harmony comes before political and civic rights (Ashworth *et al.* 2013).

If liberalism has long wanted to be thought of as a political tradition committed to freedom, a close and more critical reading suggests it is actually preoccupied with order and security.

This preoccupation with order means those who threaten 'our' security, be it student protesters, the 'poor', the unemployed, immigrants, Muslims, young people or coloured people, are more likely to be criminalised and subjected to various forms of repression.

I outline the other two cases before offering a brief examination of the justifications used for repression by the relevant governments or authorities before presenting a critical consideration of how some disciplines, i.e. criminology, legal studies and political science, have made sense of this phenomenon. I argue that we can ill afford any complacency based on a premise that liberalism is actually committed to protecting dissent. The cases examined here make it clear that liberal polities can be as enthusiastically repressive as more obviously authoritarian regimes.

The Sussex Five, 2012–2013

On a global scale, the Sussex University protest campaign was a small-scale event. In late November 2012, several dozen students barricaded themselves in the conference centre at Sussex University in protest against plans by the vice chancellor, Michael Farthing, to outsource campus services. A spokesperson for the students argued that the ongoing 'transfer of services to the private sector' was part of a 'trend of marketisation in higher education'. All the occupiers voluntarily left the occupied buildings on 4 December 2012 after the university obtained a court injunction.

Students at Sussex already had a well-established reputation for dissent. In 2010, six students had been suspended for their 'leading roles' in a student and staff campaign opposed to management cuts to teaching and the restructuring of internal democratic procedures. Senior managers called riot police onto the campus in March 2010 to break up a peaceful occupation because Registrar John Duffy claimed – falsely – that he was being held 'hostage' by students. In 2010–11, Sussex University management obtained two high-court injunctions banning all forms of protest on campus, while a total of 11 students had been suspended for protesting (Marotta 2014).

Student dissent at Sussex University in 2012 was part of a much wider campaign protesting persistent efforts by successive British governments from the early 1980s to force the 'privatisation' of universities by cutting university funding (Watts 2017). The raising of the annual cap on tuition fees to GBP 9,000 by the Cameron coalition government prompted major student activism in England during 2010. As Bailey (2016; also 2014) notes, 2011–12 was marked by the highest level of visible dissent in the UK since the 1970s. As Bailey commented, these protests were chiefly the work of students, workers and anti-Austerity activists like UK Uncut.

On 6 December 2012, two days after peacefully ending their occupation, five students had their enrolment suspended by Vice Chancellor Farthing. The vice chancellor's letter – improbably – claimed that the occupation was 'a potential hazard to sustaining the university's policies on health and safety' constituting 'a threat to the safety or well-being of students, staff or visitors'. The suspension provoked significant adverse commentary by media and politicians as an attack on freedom of speech. On 17 January, the Sussex Five were brought to a disciplinary hearing. John Duffy, the registrar at Sussex, said:

> We fully support students' rights to protest lawfully. But [given] persistent disruption, we feel we need to go further to ensure there is no repeat of the appalling behaviour that has characterised these events. We have excluded these students to protect the interests of all of the students, staff and visitors who are entitled to use the campus without fear of intimidation and serious disruption.
>
> (cited in Marotta 2014)

Adriano Marotta, one of the Sussex Five, simply said that, 'The university is criminalising our only existing avenue for dialogue. They are criminalising our right to protest' (Marotta 2014).

The Sussex Five were represented by Geoffrey Robertson QC, arguably the best known human rights lawyer in the world. The hearing collapsed following interrogation by Robertson forcing the vice chancellor-appointed chair of the panel to acknowledge his bias against the five students. The subsequent hearing denied students the right to legal representation, while reducing the severity of the allegations and staged another hearing on 11 March. The five students did not appear and were given a caution *in absentia* because of 'conduct injurious to the academic or administrative activities of the university'.

A subsequent investigation by the university ombudsman found that the university had not adopted fair procedures when it suspended the students, nor was the decision 'reasonable in the circumstances'. For the stress and hardship, the ombudsman recommended the university award a modest monetary compensation and a written apology. The Sussex Five received between GBP 2,000 and 2,500 compensation from the university as well as a written apology. The privatisation processes set loose by the vice chancellor were not affected.

If the Sussex case was a storm in a teacup, Québec's 'Maple Spring' was a full-blown hurricane which attracted global attention.

'Maple Spring', Québec

The longest student strike in the history of Québec and Canada began on 13 February 2012 and ended on 7 September 2012 (Simard 2014). The strike was prompted by university and college tuition fee increases. On 18 March 2011, Jean Charest's Liberal Party Québec government announced increased university tuition fees beginning in 2012. The increase of CAD 1,625 was a 75 per cent increase from CAD 2,168 to CAD 3,793. By December 2011, more than 60,000 students had formed a coalition called CLASSE (*Coalition large de l'Association pour une solidarité syndicale étudiante*) ['broad coalition of the Association for Student Union Solidarity'] with the aim of organising a general strike (Bonenfant *et al.* 2013).

On 13 February 2012, the majority of student associations in Québec began boycotting classes. By March, close to 310,000 students (out of 400,000 in Québec) were on strike in what was being called the 'Maple Spring' (Dupuis-Déri 2014). On 22 March 2012, a demonstration involving 200,000 people gathered in Montréal. By April 2012, Québec was in a full-blown political crisis. As Peñafiel and Doran (2017) show, the students used a sophisticated set of political, expressive and discursive techniques involving a mix of the *carnivalesque* and humour to win popular support. However, on 24 April 2012, the government excluded CLASSE from discussions on the fee increase, triggering major demonstrations every night in Montréal's streets. These protests saw mass arrests and police brutality. In protests in Victoriaville for example, 11 people were hospitalised when police fired rubber bullets and tear gas at a crowd protesting the convention of the governing Liberal party. One demonstrator lost an eye, and another suffered critical injuries.

On 18 May, a special law (Bill 78 titled 'An Act to enable students to receive instruction from the postsecondary institutions they attend') was passed by the National Assembly of Québec. This measure restricted the right to demonstrate, a measure at odds with the rights and freedoms guaranteed by the Québec Charter of Human Rights and Freedoms. Bill 78 restricted freedom of assembly, protest or picketing on or near university grounds, and anywhere in Québec without prior police approval. The bill also placed restrictions upon the right of education employees to strike. It suspended the current academic term, required demonstrators to inform police of any protest route involving 50 or more people and threatened student associations with fines of up to CAD 125,000 if they disobey. Justice Minister Jean-Marc Fournier said the legion of critics condemning the law as an attack on freedom were using 'grandiloquent and excessive' words. Fournier claimed that '[The protesters] speak of freedom of association; they are silent on the right to education ... The goal of this law is to bring respect and calm back to our society after weeks of trouble' (Perreaux and Seguin 2012).

On the strike's hundredth day, an estimated crowd of up to 400,000 filled the streets of Montréal in defiance of the law. It was said to be the largest act of civil disobedience in Canadian history. The following day, another massive protest saw a harsh police response with over 500 people arrested in Montréal and another 100 arrested in Québec City.

The fall-out was severe for Charest's Liberal Party government which may have hoped to use its tough 'law 'n order' response to advantage in the Provincial elections in September 2012. However, with opinion polls showing this was not the case, it announced a ministerial freeze on raising tuition fees on 5 September 2012, leading student strikers to vote to return to class. The Liberal government subsequently lost the election of September 2012. Bill 78 itself was quickly repealed by Pauline Maurois' new minority Parti Québecois government.

These student protests in Malaysia, Sussex and Québec provide exemplary cases of modern dissent and of state responses to that dissent.

How and what should we think about this dissent and the responses of states and authorities to it? To answer this question, I address several basic theoretical issues about how liberalism is understood as a tradition of political theory and practice.

Theorising liberalism

Beiner (1998: vii) reminds us that the social sciences have a choice to make. While Beiner is discussing the options in political theory about what doing political theory (or political philosophy) involves, the problem he is pointing to extends well beyond political theory or political philosophy into the other social sciences. As Beiner says of political theory, it begins with an implicit consensus about what 'we' share as members of a liberal political order. The job of the theorist (or the philosopher) is, apparently, then to 'articulate the basis of this consensus and to raise it to theoretical explicitness' (Beiner 1998: viii). The parallel version of this in political science involves taking a category like 'politics' or in criminology taking a category like 'crime' as if these are self-evident and ontologically stable categories before carrying out various kinds of 'empirical' studies of it.

In these cases, we depend on conventional assumptions about the ways we know things, starting with definitions. One basic assumption shared by many academics is that our language use is able to, or ought to be able to, affix a singular meaning to any word–thing relationship. On this premise, words become the basis upon which any agent is able to ascribe a fixed and literal referential meaning to our categories and things. As philosophers since Wittgenstein (1953) have understood, this account of language is problematic. Meaning and reference are understood to be mutually interdependent: this is crucial if the concept is to be used in a sentence that claims to be true, i.e. the possibility of empirical reference (Bartelson 1995: 15–18). This is a representative account of language that has been subjected to devastating criticism by Rorty (1980).

A different and still minority option, is to start from the premise that the liberal consensus is neither to be taken for granted nor its validity assumed to be self-evident. The task instead is to critically think about the nature of political order as it were, from the ground up, even if this involves calling into question the opinions, beliefs and practices that inform or sustain a liberal society (Beiner 1998: vii). This critical theoretical alternative to the common sense operating in the social sciences can be applied to the ethical and political foundations of the liberal order too many take for granted, and undo the complacency attached to core conceptual categories like 'politics', 'violence' 'social order' and 'dissent'.

Steinhof (2007) after Arendt (1994) reminds us why thinking is always to be preferred to common sense. As he notes, there is a common-sense liberal consensus or tendency to treat state-sponsored violence and repression as almost always a legitimate and defensible activity, while forms of dissent like 'civil disobedience', 'protest' and 'terrorism' always fall within the paradigm of 'senseless or downright illegitimate and detestable violence'. As he insists, common sense is not necessarily a reliable standard in these matters as it can be variously 'perverted or ignorant' or, simply, profoundly wrong. 'We are', he says, 'well advised to advance philosophical analysis against common points of view and to question pre-existing socially established frameworks of discussion and prejudices instead of docilely accepting them' (2007: 1). These remarks seem pertinent to the nascent study of criminalisation by criminologists and legal scholars.

Criminalisation of dissent

Since the 1980s, key figures like Lacey (2004; 2009), Duff (2007) and Duff *et al.* (2014) have promoted the study of criminalisation, understood – as Muncie (2009: 1) puts it – as understanding how and why 'new criminals' can be 'created' through police targeting and courtesy of new criminal justice legislation. Yet, as Hogg (2007: 83) notes, criminologists and sociologists have tended to be guided by the criminal law in adopting an 'apolitical concept of crime' which has meant that limited attention has been given to both 'political crime and the political power to criminalise'. This suggests that Matza was on to something when he pointed out how criminological positivists had 'succeeded in what would seem impossible. They separated the study of crime from the workings and theory of the state' (Matza 1969: 143).

If we are to account for criminalising dissent we need to account for the more general reliance on criminalisation as part of what Garland (2009) and Muncie (2009) have called the 'punitive turn'. There is good evidence that (neo-)liberal states like America, the UK, Australia, New Zealand and South Africa have hosted a dramatic increase in 'criminalisation' by governments of all kinds of political orientations, by adopting harsh penal policies and increasing the number of criminal offences. Arguably, the US played a lead role in the development of a tough-on-crime movement, involving the transformation of the American criminal law code, law enforcement and sentencing policies. For example, Congress created around 500 new crimes a decade between 1980 and 2007. Britain embraced an even more exuberant approach to 'criminalisation': between 1988 and 1997, the Thatcher and Major governments legislated to create around 500 new crimes. From 1997 onwards, Blair's government created an astonishing 3,023 new criminal offences before 2006. Garland (2009) stresses the links between the rampant egotistic individualism valued by neo-liberalism, expressed as a lack of care for others and a willingness to punish and further exclude those 'others' who are already 'socially excluded' like people of colour, the poor, immigrants and refugees, young people and so on, who are simply expelled from their communities. Wacquant (2009) goes so far as to argue that harsh penal policies and harsh welfare policies are intrinsic to neo-liberal statecraft.

As writers like Hall and Winlow (2016) demonstrated, there are always important politico-ethical issues at stake whenever states decide to criminalise certain kinds of conduct. Yet, as they insist, this points to some fundamental problems, one of them being the instability of 'crime' itself as a category. This is what Garland (2009) meant when he questioned the grounds on which criminology might secure its own legitimacy. Garland (2009: 118) observed that 'criminology's object [i.e. crime] is not a self-generated theoretical entity or a naturally occurring phenomenon, but instead a state-defined social problem'. This proposition implicitly raises questions about the authorising grounds on which any kind of social science project relies. On a preliminary understanding, this refers to the fact that most criminologists, sociologists and political scientists define 'crime' in ways paralleling whatever their governments define as criminal. The dangers of doing this were foregrounded by Bourdieu (2014: 35):

> one of the major powers of the state is to produce and impose (especially through the school system) categories of thought that we spontaneously apply to all things of the social world including the state itself. This points to a certain risk for anyone trying to think about the state, because we face the danger of taking over or being taken over by a thought of the state, that is, of applying to the state, categories of thought produced and guaranteed by the state.
>
> (Bourdieu 2014: 35)

This prospect is suggested by the way the 'norms' constitutive of the liberal-democratic political, social and economic project are drawn on. This means that some surveys of the criminalisation process (e.g. Brown 2013), treat it in conventional ways. To the extent that a 'normative' framework is employed, it may rely on a common-sense framework in which the state is said to be dealing with 'disruptive behaviours', and political-economic marginalisation, all involving 'combinations of material deprivation, restricted opportunity, access to drugs and alcohol, conflict and violence directed against the self and others, damages self-esteem and destroys lives'.

This common-sense tendency aligns with a disposition to treat dissent or protest as problematic. From studies like Blumer (1946) onwards, there has been an emphasis on the irrationality of crowds and the deviant nature of protest. A slightly different tendency was at work in studies which assimilated protest and dissent into the larger tendencies of liberal democracy (Tilly 1997; Tilly and Tarrow 2007) and argued that social movements and protests were simply a constitutive element in the political communication processes of modern societies and talked about 'contentious politics' and 'cycles of contention'.

This is to say nothing of how, if at all, they understand or attempt to establish the legitimacy of both dissent and of state repression – an exercise that requires that we establish the normative or ethical grounds of both dissent and of state repression. This brings us to the heart of the matter, namely several problems with the ways these disciplines understand 'dissent' and categories like 'crime', and the 'political'.

The difficulty starts with the tacit acceptance of the notion that dissent involving large-scale protests or civil disobedience is a threat to liberal order and therefore is somehow illegitimate, if only because it appears to subvert the legitimacy of democratic governments and the presumption that majoritarian democratic governments have an intrinsic legitimacy (Singer 1973). As the three case studies introduced here make plain, in each case, any choice between promoting freedom or security always came down firmly on the side of security. Let Sussex University's registrar represent this position. As Duffy said, while we 'fully support students' rights to protest lawfully' the university has excluded these students 'to protect the interests of all of the students, staff and visitors who are entitled to use the campus without fear of intimidation and serious disruption' (cited in Marotta 2014). This indicates precisely why we need to ask how much liberal-democratic states really promote freedom of speech and political engagement.

We are not obliged to take the liberal perspective at face value, nor start, as so much academic social science does, by assuming an implicit consensus or legitimacy about what 'we' share as members of a liberal political order.

We do not need to assume liberalism to be either self-evidently defensible or even that it correctly describes the current political cultures and practices of self-proclaimed liberal regimes. When we try to identify the 'norms' said to be constitutive of the liberal-democratic political, social and economic project in societies like America, Britain or Australia, we will either be told by the adherents of the legal positivist tradition that the law properly has no ethical basis, nor should it, or else we will be told by utilitarians that the law is an enterprise solidly grounded in the maximisation of utility.

In the first case, we are asked to trust the fact that states make laws that this provides all the legitimacy we need. As the famous but little-noted debate between Hart and Fuller in 1958 about the 'Nazi grudge informer' case suggests, even Hart, the greatest of twentieth-century legal positivists, agreed that this was a quite indefensible case, equivalent to believing that when a gunman puts a gun to our head and issues a command, this makes the action legitimate. In the second case, we are left at best with an obsolete eighteenth-century account of human beings found in an ethical tradition that has been subjected to devastating critique by generations of philosophers (Hall and Winlow 2016). The idea that utilitarianism (or consequentialism) secures the legitimacy of the law and/or criminal justice policy has been thoroughly discredited.

To this must be added the ever-expanding critique of liberalism. Apart from classic expressions like Schmitt (1984) and Arendt's (1994: 404) there is a growing body of contemporary work by Beiner (1998), Honneth (1996), Zizek (2008) and Dyzenhaus and Poole (2015). What this body of work acknowledges is that liberal states will go to almost any length – including creating 'grey' and 'black' 'legal holes' – to promote security, even at the expense of basic rights and freedoms. When we go looking for the legitimacy of the liberal order in a deep commitment to freedom, all we find is the insistence that security always trumps freedom.

Security vs freedom

This is the burden of Neocleous' (2004; 2008) argument about the deep affinity within the liberal tradition between the ostensible commitment to freedom and the actual commitment to measures like the criminalisation of dissent that promote security while destroying freedom (Neocleous 2008).

Neocleous (2004) demonstrates how and why Hobbes and Locke early made the case that state power exists to pursue security and that the bargain implied in both of their versions of social contract theory was that people surrendered their freedom in the state of nature in return for the state guaranteeing their security in civil society. Further, not only must any appeal to security involve a specification of the fear which engenders it (as in Hobbes), but also this fear or insecurity demands the countermeasures (security) to neutralise, eliminate or constrain the person, group, object or condition which engenders insecurity. Securing is, therefore, what is done to a condition that is insecure. 'It is only because it is shaped by insecurity that security can secure' (Neocleous 2004: 18).

Neocleous (2008) argues we see in neo-liberalism a deep affinity between the apparent commitment to freedom and the seemingly contradictory commitment to measures designed to promote security that are deeply antagonistic to freedom. The freedom/security antimony is the fault line that runs deep through liberalism. Neocleous argues that by the eighteenth century the category of security had already developed an intensely political meaning focused on the state. Adam Smith, for example, refers to the 'liberty' and 'security' of individuals as if they were actually co-terminous (Smith 1982: 405, 412, 540). In the 1780s, Bentham suggests that 'a clear idea of liberty will lead us to regard it as a branch of security' (Bentham 1843: 302). As with many concepts in this period – such as 'interest' and 'independence' – 'security' underwent a semantic shift as a consequence of the evolution of a neo-liberal politics oriented to the virtues of the marketplace and a renewed focus on individuals and freedom, but now understood as supporting these political values. As Neocleous says:

> Security became the cornerstone of the liberal bourgeois mind. Liberalism's radical recoding of the politics of order in the eighteenth century turned politics into a range of 'security measures' consistent with liberal principles. The concept of security thus became the ideological guarantee of the independent and self-interested pursuit of property within bourgeois society – the guarantee of the egoism of civil society. In doing so, security became *the supreme concept of bourgeois society* [my emphasis].
>
> (Neocleous 2004: 13)

Conclusion

This outline of some basic conceptual and politico-ethical problems with contemporary liberalism should prompt new lines of enquiry. Liberalism is not quite what it purports to be. While it appears to stress the value of freedom, it is actually preoccupied with security against threats, real or imagined. Faced with student protest, governments and university managers are prepared to criminalise dissent. While it looks as if liberal polities value the rule of law, constitutional principles and the protection of human rights, liberal states actually exhibit a well-documented proclivity to declare 'states of exception'. As Schmitt famously declared, 'sovereign is he who has the power to declare an exception' (Schmitt 1984: 1). The absence of a secure grounding of legitimacy in anything else than the use of power or force needs then to be set against the capacity of states to criminalise conduct including political activity and dissent. If sovereignty is all, and crime anything a state declares it to be, how can we proceed in ways that reinstate some ethical norms?

One such line of inquiry might pursue the notion of harm (e.g. Yar 2012). As those promoting a normative approach to the study of criminalisation have argued, the notion of harm provides a more ontologically stable basis than does the category of crime. This means we can open up questions about the respective harms associated with promoting freedom of speech and various forms of dissent (including protest, civil disobedience, whistleblowing and hacktivism) against those involving the use of surveillance, state-sponsored repression and criminalisation of dissent. This could also involve an 'ethics of recognition' (Honneth 1996) and a radicalisation of the democracy project (Zizek 2008).

As Yar (2012) argues, Honneth's theory of recognition can ground a theory of social harms. It does this because it establishes the basic capabilities associated with human flourishing while specifying the harms that various kinds of violence perpetrated by people, communities and states can cause. Yet, as Hall (2012: 18) argues, this will require more democracy, not less. The mutual recognition that Honneth and Yar seek cannot be realised until powerful interests and institutions are required to get some ethically binding authority from the majority. This can only happen when the majority are restored to a position where what they say has to be acknowledged by the elite. This requires a substantial commitment to radicalising democracy, a project in which dissent will be required to play a major and persistent role.

References

Arendt, H. (1994). *Essays in understanding*. New York: Harcourt.

Ashworth, A., Zedner, L. and Tomlin, P. (eds). (2013). *Prevention and the limits of the criminal law*. Oxford: Oxford University Press.

Bailey, D. (2014). Contending the crisis: what role for extra-parliamentary British politics? *British Politics*, 9(1): 68–92.

Bailey, D. (2016). Hard evidence: This is the age of dissent – and there's much more to come. *The Conversation*. 11 January. Available from http://theconversation.com/hard-evidence-this-is-the-age-of-dissent-and-theres-much-more-to-come-52871 [accessed 6 June 2016].

Bartelson, J. (1995). *A genealogy of sovereignty*. Cambridge: Cambridge University Press.

Beiner, R. (1998). Foreword in D. Dyzenhaus (ed.), *Law as politics: Carl Schmitt's critique of liberalism*. Durham: Duke University Press: vii-ix.

Bentham, J. (1843). *The works of Jeremy Bentham (Principles of morals and legislation, fragment on government, civil code, penal law)*. New York: Liberty Press.

Blumer, H. (1946). *Collective behaviour: new outlines of the principles of sociology*. New York: Barnes & Noble: 165–220.

Bonenfant, M., Glinoer, A. and Lapointe, M-E. (2013). *Printemps québécois: une anthologieaude*. Montréal: Écosociété.

Bourdieu, P. (2014). *On the state: Lectures at the College de France, 1989–1992*. (Trans. D. Fernbach). Cambridge: Polity.

Brabazon, H. (2006). Protecting whose security? Anti-terrorism legislation and the criminalization of dissent. YCIS Working Paper 43, Toronto, York University.

Brown, D. (2013). Criminalisation and normative theory. *Current Issues in Criminal Justice*, 25(2): 605–626.

Duff, A. (2007). *Answering for crime: responsibility and liability in the criminal law*. Oxford: Hart Publishing.

Duff, A., Farmer, L., Marshall, S., Renzo, M. and Tadros, V. (eds). (2014). *Criminalization: the political morality of the law*. Oxford: Oxford University Press.

Dupuis-Déri, F. (2014). Émergence de la notion de 'profilage politique': répression policière et mouvements sociaux au Québec. *Politique et Sociétés*, 33(3): 31–56.

Dyzenhaus, D. (1997). *Legality and legitimacy: Carl Schmitt, Hans Kelsen and Hermann Heller in Weimar*. Oxford: Oxford University Press.

Dyzenhaus, D. and Poole, T. (eds). (2015). *Law, liberty and the state: Oakeshott, Hayek and Schmitt on the rule of law*. Cambridge: Cambridge University Press.

Garland, D. (2009). Disciplining criminology. *Sistema Penal & Violencia*, 1(1): 114–121.

Hall, S. (2012). *Theorizing crime and deviance: A new perspective*. London: Sage.

Hall, S. and Winlow, S. (eds). (2012). *New directions in criminological theory*. London: Routledge.

Hall, S. and Winlow, S. (2016). *Revitalizing criminology: towards a new ultra realism*. London: Routledge.

Hogg, R. (2007). Criminology, crime and politics before and after 9/11. *Australia and New Zealand Journal of Criminology*, 40(1): 83–105.

Honneth, A. (1996). *The struggle for recognition: The moral grammar of social conflicts.* Cambridge: Polity.

Koskenniemi, M. (2012). *From apology to utopia: The structure of international legal argument.* Cambridge: Cambridge University Press.

Kumar, K. and Palansamy, Y. (2016). 'As #TangkapMO1 rally kicks off, cops warn protesters to disperse'. Royal Mail Online 27 August. Available from www.themalaymailonline.com/malaysia/article/tangkapmo1-rally-kicks-off-with-songs-speeches#sthash.QrVbj4Pt.dpuf [accessed 12 November 2016].

Lacey, N. (2004). Criminalisation as regulation: The role of criminal law. In: C. Parker, C. Scott, N. Lacey and J. Braithwaite (eds), *Regulating law.* Oxford: Oxford University Press, 144–167.

Lacey, N. (2009). Historicising criminalisation: conceptual and empirical issues. *Modern Law Review*, 72(6): 936–960.

Loh Kok Wah, F. and Boo Tiek, K. (eds). (2013). *Democracy in Malaysia: Discourses and practices.* London: Routledge.

Loughlin, M. (2015). Nomos. In: D. Dyzenhaus and T. Poole (eds), *Law liberty and the state: Oakeshott, Hayek and Schmitt on the rule of law.* Cambridge: Cambridge University Press, 65–95.

Mahriti, A. (2017). The international fallout from Najib's 1MDB scandal. *East Asia Forum.* 4 January. Available from www.eastasiaforum.org/2017/01/04/the-international-fallout-from-najibs-1mdb-scandal/ [accessed 21 February 2017].

Marotta, A. (2014). My Sussex suspension: why vice-chancellors are clamping down on protest. *Ceasefire.* 17 January. Available from https://ceasefiremagazine.co.uk/vice-chancellors-clamping/ [accessed 11 June 2016].

Matza, D. (1969). *Becoming deviant.* New York: Prentice-Hall.

Mueller, J. and Stewart, M. (2016). *Chasing ghosts: The policing of terrorism.* Oxford: Oxford University Press.

Muncie, J. (2009). The theory and politics of criminalisation. Center for Crime and Justice Studies. Available from www.crimeandjustice.org.uk/publications/cjm/article/theory-and-politics-criminalisation [accessed 10 November 2016].

Neocleous, M. (2004). Against security. *Radical Philosophy*, 100: 7–15.

Neocleous, M. (2008). *Critique of security.* Edinburgh: Edinburgh University Press.

Pascal, B. (1670/1995). *Pensees* (Trans. L. Krailsheimer). London: Penguin.

Peñafiel, R. and Doran, M-C. (2017). New modes of youth political action and democracy in the Americas: from the Chilean Spring to the Maple Spring. In: J. Bessant and S. Pickard (eds), *Young people re-generating politics in times of crises.* London: Palgrave Macmillan, 71–89.

Perreaux, L. and Seguin, R. (2012). 'Québec's emergency law blasted by critics'. *The Globe and Mail*, 18 May. Available from www.theglobeandmail.com/news/politics/quebecs-emergency-law-blasted-by-critics/article2437890/ [accessed 10 June 2016].

Rorty, R. (1980). *Philosophy and the mirror of nature.* Oxford: Blackwell.

Schmitt, C. (1984). *The concept of the political.* (Trans. G. Ulmen). Chicago: University of Chicago.

Schmitt, C. (2003). *The nomos of the earth in the international law of the jus publicum.* (Trans. G. Ulmen). New York: Telos.

Simard, M. (2014). *Histoire du mouvement étudiant québécois, 1956–2013: Des Trois Braves aux carrés rouge.* Laval: Presse Universite Laval.

Singer, P. (1973). *Democracy and disobedience*. Oxford: Clarendon Press.

Smith, A. (1982). *The wealth of nations*. London: Pelican.

Steinhof, U. (2007). *On the ethics of war and terrorism*. Oxford: Oxford University Press.

Tilly, C. (1997). *Popular contention in Great Britain, 1758–1834*. Cambridge: Cambridge University Press.

Tilly, C. and Tarrow, S. (2007). *Contentious politics*. Boulder: Paradigm.

Tuck, R. (2016). *The sleeping sovereign: The invention of modern democracy*. Cambridge: Cambridge University Press.

Wacquant, L. (2009). *Punishing the poor: The neoliberal government of social insecurity*. Durham: Duke University Press.

Watts, R. (2017). *Public universities, managerialism and the value of higher education*. London: Palgrave.

Wittgenstein, L. (1953). *Philosophical investigations*. Oxford: Blackwell.

Yar, M. (2012). Critical criminology: critical theory and social harm. In: S. Hall and S. Winlow (eds), *New directions in criminological theory*. London: Routledge.

Zizek, S. (2008). *In defence of lost causes*. London: Verso.

Part II

Youthful protest and repressive law

Chapter 3

Panic works

The 'Gag Law' and the unruly youth in Spain

Kerman Calvo and Martín Portos

Introduction

This chapter discusses the role of Spain's so-called *Ley Mordaza* [Gag Law] as the backbone of an emerging regime of governance of young people, defined by securitisation, new forms of policing and soft repression together with surveillance. Properly referred to as the *Ley Orgánica 4/2015, de 30 de marzo, de protección de la seguridad ciudadana*, the law has been at the centre of much controversy. In line with the generalised 'punitive turn and increases in criminalisation on the part of governments of all political orientation' (Bessant 2017: 206), criminal law experts as well as social movement scholars have criticised this legal change for contributing to criminalising political dissent. Amnesty International and other human rights organisations have claimed that the law poses a 'threat to human rights'.[1] A particular focus of attention has been the law's inclusion of a number of protest activities (often associated with the 15M movement) as punishable offences. Under the new legal framework, 'alterations of public peace' are qualified as more severe offences if they take place in the context of demonstrations (Comisión Legal Sol 2015: 111).[2] The Gag Law allows authorities to perform indiscriminate personal identity checks and impose discretionary penalties on some forms of political dissent; these include the so-called *escraches*, but also the organisation of protest activities within the premises of bank branches or the permanent occupation of public spaces for the purpose of protesting.[3] Fines are particularly harsh in relation to unauthorised protest actions near parliament buildings. From July 2015 to January 2016, under the Gag Law framework, the police filed more than 6,200 penalties under allegations of 'disrespect'.[4]

This chapter draws on critical criminology and Marxist social legal thinking to analyse the role of criminal law as an instrument of social control. The Gag Law is the central element of a new form of governing young people in times of austerity, globalisation and insecurity. In this framework, criminal law hardly takes a neutral approach to social and political conflicts. Dominant social groups use it not only to shape and restrict particular forms of behaviour, but also to define and enforce dominant views about what is morally and politically acceptable and what is not (Garland 1990; Wacquant 2009). According to this approach, the dramatic rise in imprisonment rates across democracies (including Spain) is the consequence of harsher criminal laws that, despite decreasing actual crime rates, are used to mitigate the anxieties caused by social change. In particular, the new face of punishment is related to rapid economic and technological change, societal individualisation, globalisation, flows of migrants and labour precariousness. Young people are especially affected by these transformations.

The concept of *securitisation* plays a key part in this conceptual framework. Securitisation refers to:

> the positioning through speech acts (usually by a political leader) of a particular issue as a threat to survival, which in turn (with the consent of the relevant constituency) enables emergency measures and the suspension of 'normal politics' in dealing with that issue.
>
> (Buzan *et al.* 1998; McDonald 2008: 567)

Originating from the international relations literature, the use of this concept quickly spread to other fields, including migration and refugee studies, but also to accounts of contemporary panics and mass terrorism (Castelli Gattinara and Morales 2017). Securitisation and punishment are intertwined: the legitimacy of criminal punishment depends on perceptions of the targeted population as a serious threat. While those perceptions might be grounded on objective claims, ideas of risk and danger can also be inspired by a project to install new forms of governance where the targeted group is subjected to new forms of domination.

This chapter claims that the Gag Law is the outcome of a previous process of *securitisation*, launched by the government and conservative forces, which aimed at framing young people's protest activities against austerity as a threat to the security of the Spanish state. The chapter situates the legislation as part of an expanding repertoire of technologies of discipline, control and surveillance that reframes societal responses to economic uncertainty, labour precariousness and a growing feeling of social injustice as security problems. This includes new forms of policing, the introduction of new administrative restrictions against the organisation of protest activities at the local level and the tightening of dissent surveillance.

The chapter begins by substantiating the process of securitisation. The securitisation of protest movements is a complex 'relational struggle over threat-making and meaning-making, which shapes the antagonistic actors involved' (Dunn Cavelty and Jaeger 2015: 177). We connect contemporary processes of regulation and the control of youth with the recent cycle of social mobilisation against austerity. The chapter then moves to explaining securitisation. This phenomenon is presented as a response by conservative elites to what they perceive as unruly youngsters. Different actors participate in a process wherein speech practices are deployed alongside other strategies for the production of meaning, including practices of security and forms of soft repression (Martín García 2014; Oliver Olmo and Urda Lozano 2015).

Context, however, is not irrelevant to the success of securitisation. The next section explains the specific configuration of the structure of political opportunities in Spain that greatly facilitates the activation of securitisation as a strategic resource in the hands of lawmakers. The introduction of new forms of governance for the Spanish youth draws on societal fears against young people in general and young offenders in particular. These fears, which hardly correlate with actual crime rates, give legitimacy not only to restrictive policies in the fields of civil liberties and human rights, but also to societal definitions of young protesters as menaces to be feared. In the concluding section, we highlight the main arguments of the chapter and signal some avenues of further inquiry.

New forms of governing young people in the shadow of the Great Recession

Framing dissent as a threat is not a recent development. The literature on the Spanish squatter and decent housing movements emphasises how different governments have insisted for years on the risks associated with social movements demanding a full reconsideration of property rights (Aguilar and Fernández 2010). Securitisation, however, has been overhauled after the terrorist attacks in the United States in 2001 – in line with a generalised turn towards punitivism in most Western democracies. Aided by highly developed surveillance technologies, and in a context of growing social panics, some governments are becoming too prone to cataloguing not just migrants or refugees, but also legitimate forms of dissent as a fundamental threat.

This questions the principles that inspire democratic forms of governance. We contend that the 2015 Gag Law lies at the heart of a new form of governing mobilised young people in Spain. Instead of embracing 15M campaigns as a sound vehicle for the expression of political opinion by previously disenfranchised constituencies, Spanish conservative forces have opted for repressing it, developing a set of technologies of control where securitisation, soft repression and surveillance play a key role.

Securitisation is a dynamic and interactive process whereby opposing sides compete to influence public perceptions on salient issues. Securitising agents, typically politicians, policy- and/or opinion-makers, shape the public's view on some social groups. Those subjected to securitisation often defend themselves by advancing alternative discourses, normally accusing securitising agents of manipulation. Power, however, is not evenly distributed: 'securitised' groups hardly have the resources to mount a robust defence at the level of public representations and mass discourses. Securitisation always has a purpose; it aims at displacing securitised groups aside from the centre of political power, justifying new forms of governance based on tighter control and possibly intrusive surveillance. Speech practices lie at the heart of securitisation. Simply put, the general public might change its mind about a certain group when certain agents, whose legitimacy to speak about public affairs is undisputed, relentlessly *say* that that group poses a threat.

In what terms does the Spanish conservative milieu speak about young people? Politicians and conservative opinion-makers distinguish – often subtly – between 'good' and 'unruly' youngsters. While the latter are those who take to the streets and fight, the former are those who cope with the crisis through individualist, silent means. Conservative Prime Minister Mariano Rajoy, whose government has implemented harsh austerity measures since late 2011, paid homage in 2012 to the 'overwhelming majority of Spaniards who do not protest against governmental decisions, accept sacrifices and work hard to let the country overcome the crisis'.[5] PM Rajoy's views, since then assumed by the ruling Partido Popular (PP), defined austerity policies as beneficial to the country in the long run, while accusing challengers of not pursuing the common good. Acts of resistance to austerity were presented as systemic threats, strategies that were not guided by positive goals but, instead, by a will to disrupt normal politics.

This reaction is linked to the mobilisation of popular dissent during the recession's nadir. Following a street march that was met with repressive actions by police, around 130,000 people moved on to occupy Madrid's Plaza Puerta del Sol, the heart of the Spanish capital, on 15 May 2011.[6] Protesters contested the austerity policies implemented in the wake of the Great Recession and called for 'Real democracy now!'. Thanks to social media platforms, information on the sit-in snowballed, and protest actions escalated. Over the subsequent days, camps – organised around open, popular, grass-roots and non-hierarchical working groups – were replicated in almost 200 cities across the country and abroad, involving hundreds of thousands of participants.

The campaign built on anti-globalisation, student and precarious workers' struggles (Portos 2016). The 15M signalled a return to mass protest as the main repertoire of social protest, wherein dissenting young people played a central role. It triggered a broader cycle of collective action directed against austerity policies. Many of these activities defended specific sectors, such as housing and public services (e.g. the education and health systems). The overall wave of protest was unprecedented in the country's recent history because of its mobilisation capacity, media salience and impact on the institutional and extraconventional arenas (Romanos 2016). Drawing on Protest Event Analysis, Portos (2016) argues that the Spanish cycle of protest consisted of three stages: the pre-mobilisation period with early risers; the peak of protest that followed the 15M shock in mid-2011; and protest decline since late 2013.

Based on studies on the protest participants, we know that left-wing, socially embedded, politically interested and highly educated people were more likely to engage in the *indignados'* protests (Calvo et al. 2011; Likki 2012; Calvo 2013). Importantly, the Spanish 15M campaign and the subsequent wave of protest had an important generational component. To a large extent, mass mobilisations have been the expression of a young generation blaming politicians, financial leaders and the mainstream media for the dearth of opportunities for personal and professional fulfilment (Antentas 2015). While anti-Austerity protests were not the exclusive form of action of this youthful social movement, it has been widely acknowledged that young people were over-represented in these events: the vast majority of participants were 19–30 years old (e.g. Likki 2012; Calvo 2013). According to Antentas, the highly educated, urban and digitally native youth embodies:

> the rising instability of the middle classes and the upper echelons of the working class, and [...] is facing life prospects that are much more uncertain and fragile than what their parents faced. The indignant generation represented the young middle class with uncertain personal biographies and future perspectives.
>
> (2015: 147)

For many (especially young) people, the recession shock and the socio-economic consequences that came about were a reality check; they made clear that their hopes for social mobility were unrealistic.

The dashed expectations of a generation and frustration, a sense of injustice and righteous anger motivated many young people to participate in the protests. This argument recalls the notion of *grievances*. Grievances are exogenous shocks (e.g. unemployment, income deprivation) and the emotional and attitudinal consequences (e.g. discontent, fear, resentment) which may disrupt taken-for-granted routines and lead to mobilisation (Snow *et al.* 1998) Some exclusionary factors, such as being economically disadvantaged (with members lacking money, work or having precarious, part-time or unstable jobs) can give way to mobilisation under various circumstances (Grasso and Giugni 2016). In fact, strain deriving from worsening employment conditions fostered protest participation in Spain from 2010–2012 (Galais and Lorenzini 2017). However, grievances were not the only determinants of protest. Levels of mass contestation allowed for the shaping of public discourses around grievances and reintroduced aspects related to inequality, social justice and poverty into the political agenda (della Porta 2015).

Additionally, anti-Austerity protests embodied a fight to take control of urban space. Throughout the wave of contestation, multiple arenas of resistance were created. There were encounters between police and protesters at different sites: from the squares during the 15M occupations (fighting police attempts to force activists out in Plaça Catalunya and Puerta del Sol), to attempts to surround the Congress (*Rodea el Congreso*) in September 2012, to mobilisations in solidarity with people about to be evicted from their homes (including tactics such as sit-ins and chainings) to the urban struggles of 2014. For example, clashes between activists/residents and the police took place in the underprivileged Gamonal neighbourhood (Burgos) in light of the local government's plans to transform a boulevard for pedestrians into a parking area. Similarly, a violent outburst followed after the police forced the squatters out of the emblematic Can Vies social centre in Barcelona to demolish it.

Practices of securitisation, surveillance and soft repression

The conservative media has been particularly active in framing mobilised young people as antisocial, unruly and even barbaric. García (2016) has put together a wealth of conservative opinion pieces on the 15M movement that show the extent to which young people's protest actions triggered a hostile response by journalists and opinion-makers writing for such dailies (e.g. ABC, *La Razón*). Reflecting on the spread of mass protests since spring 2011, a well-known conservative commentator argued:

> today we should speak about an 'unsatisfied mob', a broken spoiled brat … Spain is the heaven of the mass-man, although this is a European, and probably Western phenomenon … the crisis is deep, abysmal.
>
> (Sánchez Cámara, ABC, 11 August 2011[7])

Seemingly influenced by the ideas of Le Bon (1896), who characterised 'crowds' as mere 'masses without reason', some conservative opinion-makers represented politically organised young people as 'infantile mobs'. According to José Carlos Rodríguez, a journalist at the very conservative *Libertad Digital*, the *indignados* were in reality selfishly complaining because society 'does not give them all they want. It is an infantile outrage, characteristic of an infantilized society'.[8] Instead of framing protest actions as political, anti-Austerity protesters have been portrayed as 'rowdy', 'troublemaking', 'shabby', 'savage' and 'rabble-rousing' people. At the same time, they are 'very young', 'immature' and 'lacking in good judgement and the capacity to reason' (García 2016).

This is why they are described as 'spoiled brats', 'daddy's boys' and 'kiddos'. Infantilising is a governance strategy in the classic Foucauldian way; when deprived of maturity, subjects seem more amenable to unrestricted forms of control and surveillance. Also, immaturity leaves room for compassionate responses on the part of controlling agents, who can combine hard repression with soft mechanisms of supervision and surveillance.

An exploration of how government officials explain their handling of protest performances reveals a tendency towards misrepresenting many events as violent. When asked in 2015 about whether the 15M movement had had a good influence on Spanish democracy, Concepción Dancausa, the head of security forces in the Madrid region said that, 'in any case they have helped Podemos; other than that, they have not helped any other political party or contributed to improving the situation'.[9] A couple of years before, the former head of the National Police had accused the 15M activists of 'kidnapping cities'.[10] Conservative politicians tried to define a new problem, that of selfish and unruly youth, which was sabotaging the country's efforts to get over the deep recession. This laid the ground for a harsher discourse on criminal protesting, and more generally for a new system of governance based on stringent prohibitions, tough policing and close surveillance.

The government's handling of the so-called 2012 'Valencian spring' [*Primavera Valenciana*] illustrates how government officials and police authorities both accused activists of inciting fear and fostering crime. Scores of students occupied a high school to contest budget cuts in public education. Amongst other claims, they accused the PP-led regional government in Valencia of denying students heating in the cold winter months. The police forced occupiers out of the school premises, using disproportionate violence. According to the *Sunday Morning Herald*, 'riot cops on Monday charged and beat demonstrators, leaving several bleeding and arresting dozens of people including several minors', and then the conflict escalated.[11] The head of police forces in Valencia used military language to explain the police's brutal approach: according to him, the police 'responds when it is attacked'. Similarly, he declined to give information on the number of policemen deployed, 'as one does not give out information to the enemy'.[12]

As mentioned previously, securitisation is a communication strategy aimed at shaping the public's perception about a given object (Wæver 1995). An issue becomes such, not because something constitutes an objective threat to the state (or another referent object), but because actors define something as an existential threat to the survival of a political community. This can be pursued through verbal message repetition, but also by deploying security measures. For instance, district attorneys [*fiscales*] actively collaborated in a communication strategy that demonised protesters (Camps Calvet and García Berrio 2015: 62). They issued petitions of pre-trial imprisonment for activists detained by police. District attorneys have also (unsuccessfully) requested permanent bans on future demonstrations, together with extraordinarily harsh prison sentences. The 'Alfon case' has been particularly notorious in the Spanish context. A 21-year-old man was put in jail with no trial for 56 days and subjected to a special juridical regime that is supposed to survey potentially troublesome criminals.[13] Confronted with a process allegedly full of irregular procedures, Alfon was eventually put on trial due to charges of alleged illegal possession of explosives during the November 2012 general strike, and was sentenced to jail for four years in 2015 (Comisión Legal Sol 2015: 83–84).

Policing tactics have contributed to public representations of collective action as triggers of social disorder. In the first place, police authorities have intensified their involvement in protest, deploying new ways of pre-empting, controlling and repressing protesters. Police forces have become better equipped for surveillance and control, better organised and more acquainted with new strategies, protocols and tactics. In the context of austerity policies, the restrictive 2013 National Budget foresaw an astronomical 1,780 per cent increase in anti-riot equipment relative to 2012 (Martín García 2014: 304).[14] This allowed for the reinforcement of specialised anti-riot units and the purchase of more sophisticated equipment. For instance, the so-called *Brigada Móvil* of the regional Catalan police (a specialised unit in charge of protest policing) increased its staff from 275 policemen in 2008 to 496 in 2013 (Camps Calvet and García Berrio 2015: 56).

Spanish 'riot' police have increasingly used the controversial *kettling* tactic, which involves corralling protesters through cordons of policemen and allowing them to use only one possible exit, which often results in the detention of ordinary bystanders. Individual accounts on clashes between activists and the police highlight what is perceived as a disproportionate deployment of police forces, even for small-scale protest events. In May 2013, for instance, a national feminist *escrache* was organised in front of the PP's headquarters in several cities against a draft bill to reform abortion legislation. What the press defined as a 'sizeable deployment of police forces',[15] according to a protester, consisted of 20 police vans to merely guard the entrance of the party's headquarters in Madrid (Mato Gómez 2015: 89).

Moreover, Spanish police forces seem to be more prone to repressing protest. According to official statistics from the Ministry of Home Affairs, police only resorted to force in 0.08 per cent of events from 2013–2015 (Romanos 2016: 137). These figures, however, are not strictly in line with the information on violent clashes between activists and the police provided by the Spanish media. For instance, police forces tried to force occupiers out of Plaça Catalunya in Barcelona on 27 May 2011. The deployment of 450 policemen led to a brutal intervention during which 84 people were seriously injured, and more than 100 people were injured to various degrees (for additional examples, see Comisión Legal Sol 2015: 122–127; Oliver Olmo and Urda Lozano 2015: 93–95). According to the Comisión Legal Sol (2015: 122), there were at least 646 arrests related to protest performances between May 2011 and March 2015 – but the total state-wide number might be underestimated, representing merely 20 per cent of the total figures.

The widespread condemnation of high-scale violence led police strategists to embrace softer, yet potentially more effective forms of surveillance, control and other pre-emptive actions. Activists often talk about learning or trial-and-error processes whereby the police try to avoid being shamed by the media, while also succeeding in limiting protest (for a similar discussion on the British case, see Pickard 2014). Amongst the more indirect, incentive-based and less visible mechanisms, conducting indiscriminate personal identity checks and fining participants in contentious performances can be highlighted. The latter is commonly referred to as *bureau-repression* (Oliver Olmo and Urda Lozano 2015). While indiscriminate checks make individuals more vulnerable as fears related to lost anonymity arise, administrative repression targets a population that is particularly vulnerable to economic sanctions in a context of recession and increasing deprivation. According to some estimates, between May 2011 and December 2012, protesters received 962 fines with a total value of at least EUR

300,000 (Oliver Olmo and Urda Lozano 2015: 77–78). This accelerated in March 2013, 'when the Ministry of the Interior informed all police stations in Spain of "mandatory instructions", according to which the police had to fine anyone participating in *escraches* involving politicians' (Martín García 2014: 305). Bureau-repression is in many ways a very effective tool for containing protest; it forces activists to divert resources away from activism, while at the same time hindering 'solidarity between those who have been fined and other citizens' (Martín García 2014: 305).

Additionally, authorities have deployed intense surveillance activities. Only in downtown Madrid, the Municipal Police had 147 surveillance cameras under operation in 2014, 'forming an entire mesh of digitised control through which the panoptic gaze is imposed as an element of urban space, under the declared aim to produce safe spaces' (Ruiz Chasco 2014: 302). One-third of such devices are located in the Lavapiés neighbourhood, one of the hotspots for popular mobilisation in the Spanish capital. Importantly, protesters and their milieus' actions have been closely monitored.

Some key activists of the 15M, as well as anti-Austerity protesters more broadly, were specifically targeted by the state authorities. In early 2012, representatives of the Ministry of Home Affairs explicitly claimed that 'keeping public order' was a top priority in light of a foreseen increase in disruptive activities.[16] For instance, the police's information unit, *Brigada de Información de la Policía Nacional*, was in charge of monitoring and controlling activities of some of the movement leaders.[17] This special unit is also in charge of policing terrorist groups such as ETA, and overseeing other anti-system, neo-Nazi and Latino street gangs' activities. Additionally, several citizens' assemblies and grass-roots community centres have allegedly been subjected to surveilling actions on the side of law enforcement agencies. Although official authorities have often been reluctant to publicly recognise it, activists from the Casablanca and Maravillas collectives (and other initiatives in Madrid, such as La Traba, La Morada and La Casika) have suffered continuous eviction attempts – moreover, the realisation of their activities has been made difficult due to authorities' pressure.[18]

The opportunities for repression

The forms of administrative repression advanced since 2011 and transformations in the securitisation of protest are not new phenomena, but they are facilitated by specific configurations of the political opportunity structure. The contemporary securitisation of juvenile protesters draws on a solid master frame on security that legitimates extraordinary state action to make Spaniards feel safe. Influenced by a hard and prolonged fight against different forms of terrorism, state actors and security forces in Spain have gained further room for manoeuvring, allowing them to frame various forms of conflict as security threats and to push for more repressive forms of discipline and control.

For one thing, Spanish criminal law currently stands amongst the most severe in Europe (Brandariz García 2014). As in other democracies, the Spanish penal code has been adjusted several times since the mid-1990s, with reforms that have both incorporated a wide array of new offences and extended the length of imprisonment for most crimes. Judges have abandoned the use of alternative forms of punishment, facilitating an extraordinary increase of the prison population in the country: from 44,197 incarcerated people in 1999 to 76,079 in 2009 (Ministerio del Interior 2015: 526–527).[19] Security forces in Spain have also grown larger. While the size of police forces in Spain doubled between 2008 and 2013, it remained steady – or decreased – in neighbouring countries, such as France and Portugal. The disproportionate size of Spanish police forces is noticeable from a comparative vantage point: according to Eurostat data, while Spain had more than 520 police officers per 100,000 inhabitants in 2015, Germany had less than 300 and France only 172. These figures are not justified by homicide and theft rates, which are remarkably low – the 2014 homicide rate was 0.69 per 100,000 habitants in Spain, 1.20 in France and 0.80 in Germany.[20]

What explains the push for securitisation and punishment when actual crime rates are very low? The answer lies in the fear of crime (Zuloaga Lojo 2014). Paradoxically, many people *feel* insecure in Spain. While Spaniards do not tend to see 'insecurity' as a serious national problem – when ranking a list of problems that include unemployment and corruption – they agree with statements that portray current social life as more exposed to different forms of crime. Levels of fear of violent crime in Spain are slightly higher than those in the United Kingdom (Jackson and Kuha 2014). In sharp contrast to these views, crime rates are much higher in the UK. While there were 35.7 assaults per 100,000 Spaniards in 2013, in England and Wales, the proportion amounted to 566.5.[21] While there was one intentional homicide per 100,000 habitants in the whole UK, in Spain there was 0.65. While there were 439 thefts per 100,000 Spaniards, this ratio was 2,470 in England and Wales.[22] In a seminal work on the policing of crisis, Stuart Hall *et al.* argued that 'everything depends on how the crime statistics are interpreted [by the police], and then on how these interpretations are reported [in the media]' (2013: 14). Students of crime and punishment blame the media and political parties for promoting fear: despite the evolution of objective crime rates, feelings of insecurity correlate with both the political and the media agendas. Large social groups feel more unsafe the more mass media focus on fearful stories, violent crime and hand-picked menaces.

Fear of crime favours securitisation; it legitimises public policies that promise new security mechanisms and state budgets that are more beneficial to law enforcement and intelligence units. Additionally, securitising dissent and protest action involving young people becomes more feasible the more a society is concerned with unruly youth and juvenile crime. According to Fernández Molina and Tarancón Gómez (2010: 9), an astonishing 88 per cent of surveyed people perceived that Spanish youngsters are more disrespectful these days than 20 years ago. While 40 per cent of Spaniards were very worried about crimes committed by people in their twenties,[23] 36 per cent were worried, and only 22 per cent of respondents seemed rather unpreoccupied (ESS 2008). Importantly, a 2006 bill reformed the legislation on juvenile crime in light of the 'considerable increase in crimes committed by people who are underage, which has increased social awareness', and claimed as well that 'law enforcement has its credibility undermined due to perceptions of impunity of the quotidian and most frequent infractions'.[24] Contradicting the arguments above, data from the Home Ministry for 2004–2006 show that overall youth crime decreased – although a subset, that of violent crimes, rose (Montero Hernanz 2010: 15).

Unable to cope with new forms of social conflict – both PP- and Partido Socialista Obrero Español (PSOE)-led – governments have stimulated panics on the basis of unsound grounds (Zuloaga Lojo 2014). This kind of narrative – and legislative changes enacted, such as the 2015 Gag Law – are more likely to succeed in contexts with high levels of juvenile resilience and dissent mobilisation, where youth activism could be perceived as representing yet another form of dangerous and potentially harmful behaviour. An alliance between conservative media, Tory politicians and the police propelled a social panic about *mugging* in England during the 1970s. It allowed policymakers to focus on a new form of social menace (antisocial behaviour by unruly youngsters) as a means of crystallising anxiety by groups that were fearful of multiculturalism, globalisation and uncertainty (Hall *et al.* 2013). Fear of crime in Spain could be observed along similar lines: fearing crime in general, and youth in particular, emerges as an arena of state action where different actors find opportunities and constraints to develop restrictive policies, including those related to protest and protesters.

Conclusion

Drawing on time-honoured fears of tumultuous crowds, a major legislative change was enacted in Spain in 2015 to constrain and restrict the right to protest, the 'Law for the Protection of Citizen Safety' (also known as the 'Gag Law'). While the draft bill was presented in 2013 and approved in 2015 – despite concerns and criticisms raised by human rights organisations – the law was passed with the sole support of the ruling PP.[25] This piece of legislation was not only contested in the domestic arena, but it was also controversial in the international sphere – a *New York Times* editorial described it as 'ominous'.[26] A number of anti-Gag Law protest campaigns unfolded across the country, which combined traditional and innovative repertoires of action (e.g. petitioning; marching, including the world's first-ever demonstration by holograms;[27] and theatrical performances). The citizen platform 'We are Not Crime' [*No somos Delito*] co-ordinated much of this dissent from 2013 to 2015.

The legislative change was enacted in part as a response to the preceding cycle of popular contestation against austerity policies, where the (often precarious) youth was over-represented. In order to place popular dissent under control, authorities deployed a mix of coercive forms of repression, which evolved towards more subtle, softer tactics (including penalties and identity checks) and surveillance activities. The Gag Law represents a long-term agenda launched by the conservative milieu, which aimed at securitising juvenile protest. Although Spain is a country with low criminality rates, fears of crime are disproportionately high. Contrary to what data suggest, the government and media often depicted (young) people who engaged in both traditional and novel forms of protest as a serious threat to social order. Conservative narratives based on stimulating panics resonated with many citizens, contributing to monitoring citizens' behaviour and limiting their right to protest and dissent.

Notes

1 See www.es.amnesty.org/actua/acciones/espana-ley-seguridad-ciudadana-oct14/ [accessed 16 March 2017].
2 Comisión Legal Sol is a group of legal experts established during the 15M occupations in Madrid. They give legal support and advice to activists.
3 An *escrache* is a gathering of people that targets decision makers in order to push for a certain course of action.
4 See www.eldiario.es/sociedad/sanciones-diarias-Policia-Ley-Mordaza_0_489951750.html [accessed 8 May 2017].
5 See www.eldiario.es/politica/Rajoy-alerta-intereses-alaba-manifiesta_0_51795314.html [accessed 8 May 2017].
6 15M stands for 15 May 2011, when the mobilisation began. Activists tend to prefer this neutral label instead of other terms such as *indignados* [*the outraged*] (Romanos 2016).
7 See http://sevilla.abc.es/historico-opinion/index.asp?ff=20110811&idn=1401712023311 [accessed 16 March 2017].
8 See www.clublibertaddigital.com/ilustracion-liberal/48/rebeldes-jose-carlos-rodriguez.html [accessed 8 March 2017].
9 See www.infolibre.es/noticias/politica/2015/05/13/la_delegada_del_gobierno_madrid_sostiene_que_m_dana_marca_espana_32621_1012.html [accessed 8 March 2017].
10 See www.elmundo.es/elmundo/2012/05/11/espana/1336729955.html [accessed 3 March 2017].
11 See www.smh.com.au/business/world-business/spain-budget-cuts-spawn-valencian-spring-protests-20120221–1tme3.html [accessed 21 January 2017].
12 See http://ccaa.elpais.com/ccaa/2012/02/20/valencia/1329764951_838007.html [accessed 5 March 2017].
13 See www.eldiario.es/politica/alfon-FIES_0_78392640.html [accessed 12 February 2017].
14 See www.elmundo.es/elmundo/2012/10/30/economia/1351613307.html [accessed 3 February 2017].
15 See www.elmundo.es/elmundo/2013/05/16/espana/1368728666.html [accessed 3 March 2017].
16 See www.20minutos.es/noticia/1416360/0/policia/investiga/lideres-15-M/ [accessed 22 January 2017].
17 Ibid.
18 See http://patiomaravillas.net/2014/10/01/alertapatio; www.lavanguardia.com/local/madrid/20120919/54350581475/interior-niega-haber-investigado-a-los-lideres-del-15m-y-limita-su-vigilancia-a-los-antisistema.html [accessed 22 January 2017].
19 The Gag Law is not the first security law that the Spanish Parliament passed. The so-called *Corcuera* Bill (named after the Home Secretary who proposed it in 1992)

questioned the limits of civil rights when the government faced security threats (e.g. terrorism, the drug war).

20 See http://ec.europa.eu/eurostat/web/crime/database [accessed 14 March 2017].

21 Data come from Eurostat (http://ec.europa.eu/eurostat/statistics-explained/index.php/Archive:Crime_statistics). Figures for Scotland and Northern Ireland are smaller, though: 61.4 and 59.8, respectively.

22 Figures for Scotland and Northern Ireland are again smaller: 1,914.8 and 1,493.9, respectively.

23 This contrasts, for instance, with the French case, where only 21% of the respondents declared to be very worried about juvenile crimes.

24 See www.boe.es/boe/dias/2006/12/05/pdfs/A42700–42712.pdf [accessed 19 March 2017].

25 See www.es.amnesty.org/actua/acciones/espana-ley-seguridad-ciudadana-oct14/; www.amnesty.org/en/latest/news/2015/03/spain-two-pronged-assault-targets-rights-and-freedoms-of-spanish-citizens-migrants-and-refugees/ [accessed 14 April 2017].

26 See www.nytimes.com/2015/04/23/opinion/spains-ominous-gag-law.html [accessed 14 April 2017].

27 See www.independent.co.uk/news/world/europe/spains-hologram-protest-thousands-join-virtual-march-in-madrid-against-new-gag-law-10170650.html [accessed 14 April 2017].

References

Aguilar, S. and Fernández, A. (2010). El Movimiento por la Vivienda Digna en España o el porqué del fracaso de una protesta con amplia base social. *Revista Internacional de Sociología*, 68(3): 679–704.

Antentas, J. (2015). Spain: the indignados rebellion of 2011 in perspective. *Labor History*, 56(2): 136–160.

Bessant, J. (2017). Digital humour, gag laws and the liberal security state. In: R. Luppicini and R. Baarda (eds), *Digital media integration for participatory democracy*. Pennsylvania: IGI, 204–221.

Brandariz García, J. (2014). La evolución de la penalidad en el contexto de la Gran Recesión: La contracción del sistema penitenciario español. *Revista de Derecho penal y Criminología*, 12: 309–342.

Buzan, B., Wæver, O. and Wilde, J. de. (1998). *Security: a new framework for analysis*. Boulder, CO: Lynne Rienner.

Calvo, K. (2013). Fighting for a voice: The Spanish 15M/Indignados movement. In: C. Flesher and L. Cox (eds), *Understanding European movements: new social movements, global justice struggles, anti-austerity protest*. London: Routledge, 236–253.

Calvo, K., Gómez-Pastrana, T. and Mena L. (2011). Movimiento 15M: ¿quiénes son y qué reivindican? *Zoom Político*, 4: 4–17.

Camps Calvet, C. and García Berrio, A. (2015). La gestión neoliberal de la crisis: de la culpabilización a la represión de la protesta. In: D. Bondia, F. Daza and A. Sánchez (eds), *Defender a quien defiende. Leyes mordaza y criminalización de la protesta en el estado español*. Barcelona: Icaria, 45–66.

Castelli Gattinara, P. and Morales, L. (2017). The politicization and securitization of migration in Western Europe: public opinion, political parties and the immigration issue. In: P. Bourbeau (ed), *Handbook on migration and security*. Cheltenham: Edward Elgar, 273–295.

Comisión Legal Sol. (2015). La ciudadanía como enemiga: Balance tras cuatro años de represión de la protesta. In: D. Bondia, F. Daza and A. Sánchez (eds), *Defender a quien defiende. Leyes mordaza y criminalización de la protesta en el estado español*. Barcelona: Icaria, 107–139.

della Porta, D. (2015). *Social movements in times of austerity: bringing capitalism back into protest analysis*. Cambridge: Polity Press.

Dunn Cavelty, M. and Jaeger, M. (2015). (In)visible ghosts in the machine and the powers that bind: The relational securitization of anonymous. *International Political Sociology*, 9(2): 176–194. DOI: 10.1111/ips.12090.

ESS. (2008). Variable wr20crm: How worried about level of crime committed by people in their 20s. *European Social Survey* 4. Available from http://nesstar.ess.nsd.uib.no/webview/index.jsp?v=2&submode=variable&study=http%3A%2F%2F129.177.90.83%3A-1%2Fobj%2FfStudy%2FESS4e04.4&gs=undefined&variable=http%3A%2F%2F129.177.90.83%3A80%2Fobj%2FfVariable%2FESS4e04.4_V299&mode=documentation&top=yes [accessed 12 May 2017].

Fernández, M. and Tarancón Gómez, P. (2010). Populismo punitivo y delincuencia juvenil: mito o realidad. *Revista Electrónica de Ciencia Penal y Criminología*, 12(08): 1–25. Available from http://criminet.ugr.es/recpc/12/recpc12-08.pdf [accessed 18 March 2017].

Galais, C. and Lorenzini, J. (2017). Half a loaf is (not) better than none: how austerity-related grievances and emotions have triggered protests in Spain. *Mobilization*, 22(1): 77–95.

García, J. (2016). Nuevos movimientos, viejas explicaciones. La huella de Gustave Le Bon en el 15M. Unpublished manuscript.

Garland, D. (1990). *Punishment and modern society: a study in social theory*. Chicago: The University of Chicago Press.

Grasso, M. and Giugni, M. (2016). Protest participation and economic crisis: the conditioning role of political opportunities. *European Journal of Political Research*, 55(4): 663–680. DOI: 10.1111/1475–6765.12153.

Hall, S., Critcher, C., Jefferson, T., Clarke, J. and Roberts, B. (2013). *Policing the crisis: mugging, the state and law and order*. Second Edition. London: Palgrave Macmillan.

Jackson, J. and Kuha, J. (2014). Worry about crime in a cross-national context: a model-supported method of measurement using the European social survey. *Survey Research Methods*, 8(2): 109–125.

Le Bon, G. (1896). *The crowd: a study of the popular mind*. New York: The Macmillan Co.

Likki, T. (2012). 15M revisited: a diverse movement united for change. *Zoom Político*, 11: 1–16.

Martín García, Ó. (2014). Soft repression and the current wave of social mobilisations in Spain. *Social Movement Studies*, 13(2): 303–308.

Mato Gómez, M. (2015). Apuntes para pensar el género como elemento constitutivo de la repression. In: D. Bondia, F. Daza and A. Sánchez (eds), *Defender a quien defiende. Leyes mordaza y criminalización de la protesta en el estado español*. Barcelona: Icaria, 67–106.

McDonald, M. (2008). Securitization and the construction of security. *European Journal of International Relations*, 14(4): 563–587.

Ministerio del Interior. (2015). *Anuario Estadístico 2015*. Available from www.interior.gob.es/documents/642317/1204854/Anuario-Estadistico-2015.pdf/03be89e1-dd38-47a2-9ce8-ccdd74659741 [accessed 12 May 2017].

Montero Hernanz, T. (2010). La delincuencia juvenil en España, en Datos. *Quadernos de criminología*, 9: 14–22.

Oliver Olmo, P. and Urda Lozano, J-C. (2015). *Protesta democrática y democracia antiprotesta. Los movimientos sociales antes la represión policial y las leyes mordaza*. Pamplona-Iruña: Pamiela.

Pickard, S. (2014). Keep them kettled! Student protests, policing and anti-social behaviour. In: S. Pickard (ed.), *Anti-social behaviour in Britain. Victorian and contemporary perspectives*. London: Palgrave Macmillan, 77–91.

Portos, M. (2016). Taking to the streets in the context of austerity: a chronology of the cycle of protests in Spain, 2007–2015. *Partecipazione e Conflitto*, 9(1): 181–210.

Romanos, E. (2016). Late neoliberalism and its indignados: contention in austerity Spain. In: D. della Porta, M. Andretta, T. Fernandes, F. O'Connor, E. Romanos and M. Vogiatzoglou (eds), *Late neoliberalism and its discontents in the economic crisis. Comparing social movements in the European periphery*. Basingstoke: Palgrave Macmillan.

Ruiz Chasco, S. (2014). Videovigilancia en el centro de Madrid: ¿Hacia el panóptico electrónico? *Revista Teknokultura*, 11(2): 301–327.

Snow, D., Cress, D., Downey L. and Jones, A. (1998). Disrupting the quotidian: reconceptualizing the relationship between breakdown and the emergence of collective action. *Mobilization*, 3(1): 1–22.

Wacquant, L. (2009). *Prisons of Poverty*. Minneapolis: University of Minnesota Press.

Wæver, O. (1995). Securitization and desecuritization. In: R. Lipschutz (ed.), *On security*. New York: Columbia University Press.

Zuloaga Lojo, L. (2014). *El espejismo de la seguridad ciudadana: claves de su presencia en la agenda política*. Madrid: Catarata.

Chapter 4

Controlling dissent through security in contemporary Spain

Laura María Fernández de Mosteyrín and Pedro Limón López

The forces of security

In 2015, the Spanish Parliament passed the Public Order Regulation Act.[1] This Act greatly restricted public order, and came to be known as the Spanish *Gag Law*. The legislation mobilised numerous international human rights organisations, generated intense parliamentary debate and saw the emergence of new civil liberties movements. In 2012, Secretary of State Ignacio Ulloa, the highest Spanish home affairs authority, justified the law by arguing for the need to regulate the 'proper' use of public space. At the same time, he announced the use of audiovisual control of mass demonstrations for the surveillance of 'radicals', 'in order to discriminate between those who resist in a violent way [...] and those who appropriately use their right of assembly' (Ulloa 2012: 15–16). This law is considered to be the Spanish state's principal response to anti-Austerity mobilisation since 2011 and exemplifies the government's bid to control dissent (Blay 2013; Bondía 2015); it is also widely believed to be limiting freedom of speech and protest. We argue that this legislation was not enacted in a vacuum and that there are other initiatives limiting protest and speech; as such, it can be seen as part of a larger political process.

In the period 1999–2016, consecutive Spanish governments developed regulations to tackle behaviour regarded as disruptive and a threat to stability. These actions became part of a process that saw the state's power reinforced through security as a mechanism for regulating conflict (Fernández de Mosteyrín 2013). Over the last two decades, the Spanish state has developed strong counterterrorism apparatus that has strengthened its power. While the state has not invoked special counterterrorism powers since the 1980s[2] (Llobert-Anglí and Masferrer 2015) a strong consensual narrative developed between state and civil society in the early 2000s, resulting in the codification of new criminal laws, criminal procedures, the reform of intelligence services and new regulations for political parties (Cancio 2014). In recent years, the state's strategy became more sophisticated in its attempt to prevent *jihadism* and to counter radicalisation (Fernández de Mosteyrín and Limón 2017). Throughout this period, an increasing range of behaviour, including glorification of terrorism or diffusion of extremist ideas, came to be seen as connected to terrorism. These developments had an adverse effect on freedom of speech.[3] This chapter shows how this effort at control spread beyond 'the problem of terrorism' to state-sponsored mechanisms to govern political dissent generally (Jackson 2005).

At the same time as counterterrorism spread, an alternative strategy by the Spanish state in the field of public order has impacted on young people's politics. This programme of governance started with the labelling of particular kinds of conduct and groups as 'problematic'. They included minor antisocial behaviour – 'incivilities' – which affected 'undesirable groups' such as 'foreigners', 'the underclass' or 'radicals' (Limón 2013). Young people became a particular target of such regulatory interest. This first occurred when the state began limiting young people's leisure options and their opportunities to participate in politics in the public space, as we endeavour to demonstrate below.

A growing field of research has analysed anti-Austerity movements and their responses to neo-liberal policies (Flesher-Fominaya and Cox 2013; della Porta 2015). Social movement studies provided empirical evidence, mainly macro-comparative, documenting processes of diffusion of repertoires of collective action (Flesher-Fominaya and Montañés Jiménez 2014; Romanos 2016). There is also a significant body of research examining the state's policing of protests and innovations by law enforcement for control (Gillham 2011). The emerging field of inquiry into what is described as 'the authoritarian turn' is shedding light on transformations in criminal justice systems (Garland 2001; Wacquant 2009) and on the militarisation of law enforcement in the US, Britain and France (Hallsworth and Lea 2011; Fassin 2013; Goffman 2014). Finally, there is a growing body of research that examines the impact of counterterrorism/extremism policy on political dissent, surveillance of 'suspicious communities' and freedom of speech (Mullard 2007; Cram 2012; Jarvis and Lister 2013; Kundnani 2015). Although securitisation discourses and practices are receiving considerable attention across various disciplines, there still is, with notable exceptions (Neocleous 2000, 2011; Tarrow 2015), a lack of clear theorisation of the role of the state's power in this authoritarian transformation.

We juxtapose the analysis of counterterrorism and public order to better understand the Spanish state's response to the Indignado movement and youth politics more broadly. To this end, we move away from the categories of collective action to analyse state power. We argue that the Spanish state resorted to engaging security institutions, like the police and penal system, and deployed the rhetoric of securitisation to regulate dissent. This strategy resulted in the expansion of state powers to label and monitor specific groups, and to punish them. In short, we offer a critical review of security policy and public order regulation for the period 1999–2016, focusing on young people and the state's interest in imposing social order. We show how this happened through state action that identified and labelled particular groups of young people.[4] To this end, we develop a genealogy of state action in the period and examine recent post-15M regulations. We will highlight two different forms of state power that illustrate its capacity to designate certain behaviour as 'radical', and the power to treat certain actions as 'crime'.

The problem as security and need for social order

State power has been theorised as the ability to regulate social relations (Neocleous 2000). It can be recognised by its penetration into the everyday lives of citizens. Following Mann's concept of infrastructural power, it is the 'capacity of the state to actually penetrate civil society and to implement logistically political decisions throughout the realm' (1984: 189). We focus here on state action aimed at maintaining order inside its territory. As devices to pacify oppositional movements, we examine the role of institutions in which that power is enshrined and how those institutions have been deployed to repress certain actions in the name of social order.

State power is frequently defined in terms of its external capacity – military power – and its bureaucratic power (Soifer 2008; Enríquez and Centeno 2012). One way of understanding the expansion of this bureaucratic capacity is to focus on the political, social, legal and technical expressions of the government regulation. However, these regulations affect various groups differently. While some groups are described as 'threats to social stability', it is paramount to examine how this process occurs and how youth are affected by state action.

We concentrate on the state's use of the law to regulate behaviour. As Neocleous argues, 'we need to address the constitutive power of the state over civil society and the way this power is expressed by law and administration continuum' (2000: 11). By looking at how political practices are identified as threats, prohibited by law and policed by security institutions, we can reveal how young people's politics is being governed through security discourse. This is a disciplinary mechanism designed to prevent dissent (Neocleous 2011: 193) which operates, we claim, through surveillance, designation, regulation and punishment.

Before the state can regulate certain social practices, a process of identification takes place. Labelling is an interactive process in which powerful actors like state authorities and security operators (such as intelligence communities, law enforcement, judges, etc.) and actors in civil society (media, experts or think tanks, for example), tag groups and individuals whose behaviour is considered deviant or a threat. Deviance depends on the reaction of others and the consensus imposed by those with power (Becker 1963). Therefore, 'antisocial behaviours', 'extremism', 'radicals', even 'terrorism', are defined as disruptive before being categorised as a crime. This process is key to understanding criminal coding. As the Spanish case shows, discourses of threat and instability of certain youth activities preceded the expansion of criminal laws.

Statecraft in contemporary Spain: youth protest and security policies

While the prominence of young people as political actors has been highlighted since 15M (García Galera 2014; Lobera 2015; Subirats 2015), the Spanish state had been developing countermeasures to manage or repress young people's political practices since the early 2000s. These measures included the criminalisation of certain leisure activities, surveillance of public spaces, infiltration of demonstrations and other control strategies that, as said above, culminated in the criminalisation of dissent through the 2015 Spanish *Gag Law*.[5]

This process began when the anti-globalisation protests in Seattle at the World Trade Organization summit (1999), in Prague (2000) and Genoa (2001) were replicated in Spain through the mobilisations of movements like *Nunca Mais* and the *anti-Iraq War* mobilisation in 2002/2003 (Aguilar Fernández and Ballesteros Peña 2005). The official response came in the form of urban civic ordinances (Limón 2012, 2014). The first legal instrument used against the organisation of youth movements was the so-called *Anti-street binge drinking Law*. In 2002, the Madrid Assembly approved Law 5/2002, of 27 June, on 'addictive disorders'. This initiative tightened sanctions for the 'undue' presence of youth in public spaces, and established 'zones of priority action' at the discretion of city and regional government.[6] By including regulations for 'hygiene' and civic coexistence,[7] many municipalities restricted 'free' participation in public space, which stigmatised urban groups engaged in political action (García and Ávila 2015). These arrangements were widely backed up by surveillance to prevent 'uncivil' behaviour (Chasco 2014; Galdón-Clavell 2015).[8]

The impact of the 11 March bombing in 2004 in Madrid was transformative in many ways for state counterterrorism work, especially in the expansion of intelligence (Navarro and Bonilla 2007) and the increase in law enforcement resources (Reinares 2008). These events also transformed Spanish politics. New forms of mobilisation developed and became central to young people's political action. For the first time, new information technologies (e.g. mobile phones) entered the theatre of dissent (Flesher-Fominaya 2011).

Despite attempts by the mainstream media to criminalise protest movements such as *Nunca Mais* and the *Anti-war* movement (Aguilar Fernández and Ballesteros Peña 2005: 111–112), there was at this early stage no clear identification of 'youth' as a threat to social stability. This 'identification' began with the mobilisation of *Movimiento por la Vivienda Digna* [Movement for Dignified Housing] in 2006, which predominantly involved young leftists and student leaders who demonstrated against their limited access to affordable housing (Aguilar Fernández and Fernández Gibaja 2010: 689–696). It was in response to this mobilisation that the labelling of young people as a serious threat to social order became paramount.

The Movement for Dignified Housing focused media attention on young people's particular political identity. Through this framing process, they became associated with various 'problematic' issues such as violence, antisocial behaviour and addictions (Alcoceba and Matheus 2010). This representation of youth shifted from the idea of them being alienated from institutional politics, to them being harbingers of unconventional and dangerous political action (Benedicto and Morán 2015: 84; Espinar-Ruiz and González-Río 2015). This labelling process was then used to justify a tougher police response.

This activity peaked following the 2011 appearance of the social movement *Juventud Sin Futuro* [Youth Without Future]. Their protest crystallised on 15 May, leading to the wide-scale occupation of public spaces across the country and the emergence of the so-called 'Indignado' movement, which revealed a new political subject whose mobilisation urged multiple changes demanded by youth,[9] such as housing, a fair income, etc. The creation of new organisational structures, discourses (Abellán Bordallo 2015: 18; Jerez *et al.* 2015; Subirats 2015) and the rise of informal mechanisms of participation were evident (Lobera 2015), especially by young people (García Galera 2014). The new political repertoires included a sophisticated use of social media and the occupation of public spaces (Flesher-Fominaya 2015; Romanos 2016).

The state responded to the Indignados by discouraging young people's participation in political protest (Hueso *et al.* 2015: 62). This was achieved through the redefinition of behaviours allowed in public spaces (including tighter control of the Internet) and the surveillance of social media.[10] Additionally, most demonstrations were strongly policed (Blay 2013; Fernández de Mosteyrín 2013; Pisarello and Asens 2014; Camps and Vergés 2015) by undercover and uniformed officers.[11]

In reaction to the protests, the authorities also advanced a counter-narrative that connected acts of political dissent with extremist behaviour. The label *anti-sistema* ('extremist' or 'radical' left-wing activists) was not new, as it had appeared in the early 2000s in public debates. However, it became increasingly common in press reportage and official narratives following anti-Austerity protests.[12] This narrative of 'young political actors as a radical threat' was used to censor humour (Bessant 2017), to demand tougher criminal regulation, to delegitimise innovative repertoires by depicting protesters as terrorists and to justify policing of demonstrations.[13]

Police power, bureaucratic power and 'radical youth'

In the period 2014–2016, four legal pillars in the Spanish security system were reformed: private security, criminal law, public order policing and national security.[14] In each case, regulations expanded the definition of security, labelled new crimes and threats, prescribed the actors involved and the mechanisms of accountability. Either oriented to counter religious terrorism or to regulate public order disturbance in anti-Austerity protest, each of these aspects resulted in a limitation of speech and the restriction in the use of public spaces. Glorification of terrorism and dissemination of extremist content essentially affected young social media users, and the criminalisation of certain uses of public space affected youth political practices.

The Public Order Regulation Act was said to be justified because of a rise in criminality, a sense of insecurity and claims that citizens demanded peace and tranquillity.[15]

The increasing regulation of public space expanded the range of unlawful offences, as well as increasing the sanctions, which made this statute 'tailor-made' for use against the anti-Austerity movements. Additionally, many of the innovative repertoires that developed on or after 15M became 'an administrative offence' rather than a 'crime'.[16]

Most of the proscribed actions are common political practices amongst young people (Feixa and Nofre 2013). In particular, the category of 'passive resistance to authority' criminalises civil disobedience (art. 36). It is considered a very serious offence and is heavily fined, together with 'public order disturbance any time a meeting or demonstration is taking place before the seats of political institutions' (of the type popularised by Occupy). New 'preventive' police powers include the use of 'stop and search' to verify identity, including people whose face is not completely visible. This provision affects people who wear religious clothing and especially juveniles wearing hoodies (both pejorative symbols of potential radicalisation). Restrictions on circulation in public spaces are permitted on the basis of a preventive logic and may be accompanied by registrations and body controls (arts. 16, 17, 18) in order to search for evidence of intent to undermine public order. Police agencies are allowed to employ static and mobile camcorders (art. 22), expanding their power to monitor mass events. The creation of security zones in public spaces and the dissolution of gatherings for preventive purposes are allowed (art. 23). The Act further codifies extraordinary measures in cases resulting from the 'spontaneity' of certain forms of mobilisations (art. 21). The subject 'offender' has been expanded in a way that might make the people who convene demonstrations responsible (art. 30–32). Sanctions have stiffened.[17] The vagueness of the drafting of the law has been analysed in detail by legal experts (Amnesty International 2014, 2017), but the logic that underpins the regulation is preventive[18] – suspicion based and spatial in nature.

Further measures include the national *Criminal Code*, which was reformed in 2015.[19] It continued to broaden the definition of the crime of terrorism to include behaviours that were related to social media and the dissemination of certain ideas. It also codified attacks on authority, and transformed certain offences into crimes (e.g. graffiti or petty robbery).

The *Ley de Enjuiciamiento Criminal* [Criminal Procedure Law] is a law that regulates the rules of criminal procedure in Spain. It was modified in 2015 with the special purpose of regulating police powers to develop undercover investigation of cybercrime. The regulation has expanded police powers and legalised internet surveillance.

In all the aforementioned legislation, the scope of police power and bureaucratic power has expanded. Law enforcement agencies have gained greater discretionary power, have increased their operative capacity[20] and have increased their disciplinary and surveillance power. On the other hand, certain controls over law enforcement agencies have been removed. Police officers, for example, can directly fine specific offenders under the Public Order Act, without recourse to the courts. Finally, this same regulation penalises the recording of images of the authority, which could be argued to obstruct citizen and press control and discourage oversight.[21]

The term *Bureau-repression* alludes to 'the use of administrative sanctions to criminalise, repress, penalise and ultimately defuse protest carried out by social, political and civic movements' (Oliver and Urda 2015: 1314). This type of suppression is less visible and less violent, but more coercive. It results in bureaucratic intimidation (García and Ávila 2015) that differentially affects social groups, especially marginalised youth and anyone labelled as 'radical'.[22] In effect, it discourages political engagement and quashes political dissent.

As state power increases, so too does its ability to penetrate and monitor society. Although the measures taken to avoid criminal law in matters of public order appear to be an innovation in the state's response to anti-Austerity, this is not entirely the case. As we have argued, the shift to a soft repressive strategy by the Spanish state is a legacy from experience. The Spanish government has avoided the invocation of special powers to tackle terrorism since the 1990s. One of the most successful 'lessons learned' by security operators and authorities is that the 'the full force of the law' narrative made possible and socially acceptable a wide range of legal changes throughout the 2000s. For instance, these changes normalised the criminalisation of glorification of terrorism and the outlaw of Basque radical nationalist parties, eventually repressing dissent through a soft strategy (Fernández de Mosteyrín 2013). We can follow the development of this strategy and the merging of logics of preventing terrorism and controlling public order if we now turn briefly to the most recent security programme that especially affects young Muslims.

Controlling dissent through National Security

The National Security Law of 2016 established a new structure for the Spanish security system. Widening the concept of public security to include internal security and defence, this law incorporates law enforcement, private security and military and civil society's involvement in surveillance intended to prevent terrorism, amongst other threats. Official concern about public order has merged with counterterrorism measures as part of this framework. This law is also the context in which state-sponsored 'counter-radicalisation' is being implemented. According to the official definition, radicalisation is 'any incidence that may involve the initiation or development of a process of radicalisation or gestation of extremist, or hatred for racist, xenophobic, belief or ideológical reasons'.[23] Even though this definition refers to religious radicalisation, it is also used 'to prevent' certain communities considered 'vulnerable to be drawn into extremism'. While it specifically focuses on young Muslims, all 'politicised youth' may also be surveilled. According to National Security and law enforcement agencies, anarchism, far-left and -right organisations are considered to be evolving drivers of extremism (DSN 2013, 2014, 2015). Therefore, their members and sympathisers will also be under control and monitored. The vague official narrative of this counter-radicalisation programme means it could include a variety of social groups.

In the Spanish Counter-Radicalisation Strategy, counterterrorism and public order can be seen to be merging and, in this context, police power and bureaucratic power are especially visible. This programme is also a good indicator of Mann's concept of infrastructural power since it requires the mobilisation of all public sectors and, in particular, civil society institutions,[24] stimulating surveillance not only amongst security agencies, but also amongst sectors of society (Fernandez de Mosteyrín and Limón 2017). Under this logic, which has been extensively analysed in the UK and the US (Hickman *et al.* 2012; Kundnani 2015), the Spanish state makes use of its administrative apparatus to monitor social order and pacify dissent.[25] As in other Western countries, groups deemed to be potentially 'at risk of radicalisation' are young people – especially young Muslims – and young activists.

Conclusion

This chapter documented how Spanish governments responded to young people's dissent in the period 1999–2016.

We have offered a critical review of state security policy and public order regulation for this period by focusing on the state's interest in imposing social order. We show how this happened through state action that labelled particular groups of young people as dangerous 'radicals'. In this process, the Spanish state expanded its power through a labelling process, via increased surveillance of those named problematic and through their regulation and punishment.

The Spanish government's approach provides a strong case study that illustrates the convergence of counterterrorism and public order strategies by extending the reach of criminal law and increasing its surveillance of public spaces via the expansion of routine coercion and control, without the need to invoke special powers. In this process, legacies of past counterterrorism experience, the new cycle of anti-Austerity protest and the international expansion of jihadist violence have operated as transformative forces. This chapter has also demonstrated how this has impacted upon various kinds of youth dissent.

Notes

1 Ley Orgánica 4/205 de 30 de marzo para la Protección de la Seguridad Ciudadana. Thereafter, LOSC.
2 Spanish counterterrorism strategies in the early 1980s were tainted by allegations of a 'dirty war', resulting in a delegitimation of the new democracy in the Basque Provinces and boosting recruitment by the Basque organisation ETA. 'Lessons learned' by authorities in the 1990s later helped shape counterterrorism.
3 In 2016, the case of the Spanish musician C. Strawberry resonated internationally as he was given a sentence for 'glorifying terrorism' in the lyrics of his music.

4 Our case study draws on our previous research, the relevant scholarly literature, primary legal documentation, official speeches, parliamentary debates and press accounts.

5 In the field of counterterrorism, strategies to address 'low intensity' youth street violence were developed in the early 2000s (kale borroka) in the context of the Basque Conflict. The Criminal Law of the Minors was transformed in order to hold minors accountable. Following post-9/11 EU counterterrorist measures – especially the EU Terrorist List – juvenile organisations connected to ETA were banned in Spain. See: Fernández de Mosteyrín 2012.

6 The Spanish state is highly decentralised. While national security is centralised through law enforcement agencies (National Police Agency and Civil Guard), regions (Comunidades Autónomas) are free to develop their own legislation in many other areas. In terms of security, however, only the Basque Country and Catalonia regions have their own law enforcement agencies. Cities and towns across the country have their own local policing agencies with their own responsibilities and accountability systems.

7 Civic Ordinances adopted in the Spanish state particularly affected are: Law 5/2002 (on drug dependence and other 'addictive disorders' within the Madrid Autonomous Community – Anti-street binge drinking Law); The Metropolitan Police and Government of the Villa of Madrid of 1948 (in some situations still in force); ANM Ordinance 2009/6 (on cleaning public spaces and waste management); Law 31/2010 (of the Metropolitan Area of Barcelona); General Plan of Seguretat in Catalonia 2012–2013; Law 4/2003 (on the Ordinance of the Public Safety System of Catalonia); The Ordinance on Measures to Promote and Guarantee Citizen Coexistence in Barcelona's Public Space (adopted on 23 December 2005/amended 29 April 2011).

8 Law enforcement can use video surveillance to ensure the specific use of public spaces (Ley Orgánica 4/1997, BOE 1997, art. 1).

9 Between 2010 and 2015, mobilisations in Spain increased. In 2010, 21,941 demonstrations took place across the Spanish state. These practices intensified after the emergence of the Indignado movement, peaking in 2012 and 2013 (44,233 and 43,170 respectively).

10 In 2016, Amnesty International reported that the Spanish High Court sentenced 25 people for 'the glorification of terrorism'. In total, Operación Araña (2014–2016), arrested 73 people for this crime. This operation entailed the interception of social media messages (Amnesty International 2017). For a compelling case study on the criminalisation of humour and satire in Spain, see: Bessant (2017).

11 Undercover policing in demonstrations has been reported since 2001 anti-globalisation demonstrations in Spain. According to press accounts and magistrates' courts (Ríos *et al.* 2001), since 2011, a number of protests resulting in public order disruption have revealed the undercover policing of demonstrations. These practices have been recorded and disseminated by protesters as part of their repertoires of action. See: Rodríguez-Pina 2012.

12 As we have argued elsewhere (2012), one way authorities criminalised 'radicals' was to make a connection with the terrorist organisation ETA. In the Spanish public debate, the slogan 'todo es ETA' [everything is ETA] became part of a powerful narrative of social delegitimation of violence in the Basque Conflict (Fernández de Mosteyrín 2012). Afterwards, it also became a way to censor disruptive actions in all sorts of demonstrations.

13 In 2012, after the General Strike known as 29M, the Catalonian law enforcement agency (Mossos d'Esquadra) published a report comparing public order disturbance to low-intensity terrorism. The report employed language such as 'cells', 'columns' and 'urban guerrilla' (see: www.abc.es/20120402/local-cataluna/abci-mossos-detenidos-huelga-general-201204021909.html [accessed 11 July 2017]). The protest known as

'Rodea el Congreso' [Occupy the Congress] ended in September 2012 with 35 detentions and 64 people wounded. On 26 September 2012, to justify the heavily policed demonstration in Madrid, the Government's Delegate in Madrid blamed the 'radicals' and 'antisistema' (www.rtve.es/noticias/20120926/cifuentes-sobre-incidentes-del-25s-policia-recibio-ataque-desproporcionado/565053.shtml).

14 Organic Law 4/2015, of 30 March, on the protection of public safety; Law 5/2014 of 4 March on Private Security; Criminal Code 2015 (previously amended in 2003, 2005, 2010) and Law 36/2015 of 28 September, National Security.

15 See: Balance of Criminality, Ministry of the Interior (2013; 2014; 2015). Also, Surveys 2702/2007; 2315/1999; 2284/1998 Spanish Centre for Sociological Research (CIS, 1997; 1998; 2007). The perception of terrorism as a problem has historically been very important for Spaniards. However, according to this data, 'insecurity' has not been a major cause for concern since the 1980s. Since the end of ETA (2011), jihadist terrorism has been mentioned as one of a number of problems.

16 By decriminalising behaviours previously included in the criminal code, and considering these same actions as administrative offences, law enforcement officers have been empowered to fine protesters directly without recourse to a judicial process.

17 Very serious infringement: EUR 30,001 to 600,000 fine (e.g. unauthorised demonstrations or demonstrations in critical infrastructures). Serious infringements: EUR 601 to 30,000 fine (e.g. obstruction of authority when executing judicial decisions, such as evictions; demonstrations in front of parliament; the unauthorised use of images of agents of authority).

18 The preventive use of detention were considered by civil liberties organisations as a mode of surveillance. Between May 2011 and December 2015, the Legal Sol Commission (the Indignados' office for civil liberties) reported 646 detentions related to public order crimes (this organisation estimates that this amounts to 20% of the total detentions in the period). Of these, only two cases ended in a court sentence (Commission Legal Sol 2015: 122). Official data are not available to support this argument further. However, while 'identification for security purposes' was, according to the Spanish Ministry of Interior, in clear decline in the last decade, 2011 and 2012 show the highest rates 8,773,862 (2011); 7,585,526 (2012). The case might be that, while the general trend in 'identification' is in decline, it is more selective by profile. According to The Spanish Ombudsperson (2010), complaints about police identity checks of foreign nationals were increasing (Amnesty International 2011).

19 The code had been reformed before in 2001, 2003, 2008 and 2010.

20 The Private Security Act (2014) regulates the private sector and makes it compulsory to co-operate with law enforcement. The status of the private sector moves from a 'junior' to an equal partner in the new regulation – from subordination to co-operation. This logic also applies to the field of National Security.

21 Rigorous investigations into police excesses in recent years is absent (Amnesty International 2014).

22 According to official statistics, people sentenced for public order crimes amounted to 13,113 (2009) and 13,745 (2015) with no significant change. Around 60% of those sentenced were people younger than 35 years old. See: www.poderjudicial.es/cgpj/es/Temas/Estadistica-Judicial/Estudios-e-Informes/Justicia-Dato-a-Dato/. This is consistent with the fact that in the period 2010–2015, detentions-investigations related to alleged public order crimes fell (28,116 in 2011 to 18,626 in 2015) while offences and public order sanctions in related demonstrations increased: 561 (2010); 376 (2011); 1,722 (2012); 981 (2013); 593 (2014). Police were already using their discretional power to channel disruptive behaviour towards administrative, rather than criminal, law. Following the drafting of the Public Order Regulation (2015), these offences rose rapidly: 87,872 (2015) to 97,947 (2016). Of these, 126,115 (2016)

and 66,832 (2015) related to 'passive resistance to authority' (i.e. civil disobedience). Fines amounted to EUR 61,175,445 in 2016. These data shed light on how the law was tailor-made to counter public disorder and discourage collective action. See: https://estadisticasdecriminalidad.ses.mir.es/

23 See https://stop-radicalismos.ses.mir.es/stop/FormServlet

24 The counter-radicalisation strategy, society's effort to prevent radicalisation, and the requirement to communicate it to the authorities via intelligence agencies and online platforms implies the institutionalisation of a de facto system of horizontal surveillance, and is a maximum expression of the state's capacity to penetrate society.

25 'Local Groups for the Detection of Radicalisation' comprise members of the local law enforcement agency, social services and the educational community and are co-ordinated by the intelligence services.

References

Abellán Bordallo, J. (2015). De la red a la calle: El proceso de movilización previo a las manifestaciones del 15 de mayo. *ACME: An International E-Journal for Critical Geographies*, 2015, 14(1): 10–29.

Aguilar Fernández, S. and Ballesteros Peña, A. (2005). El modelo de proceso político a debate. Una explicación al origen y conseucuencia del movimiento social 'Nunca Mais'. *Revista Española de Investigaciones Sociológicas*, 111(05): 111–136.

Aguilar Fernández, S. and Fernández Gibaja, A. (2010). El movimiento por la vivienda digna en España o el porqué del fracaso de una protesta con amplia base social. *Revista Internacional de Sociología*, 68(3): 679–704.

Alcoceba, J. and Matheus, G. (2010). El discurso mediático sobre los jóvenes en España. *Anuario electrónico de estudios en Comunicación Social*. Disertaciones, 3(1).

Amnesty International. (2011). *Stop racism, not people. Racial profiling and immigration control in Spain*. London: Amnesty International Ltd.

Amnesty International. (2014). *El Derecho a Protestar, Amenazado*. Madrid. Available from https://doc.es.amnesty.org/cgi-bin/ai/BRSCGI/44100114.spa%20(policing%20spain_FINAL_en%20baja)?CMD=VEROBJ&MLKOB=32906041616 [accessed 27 November 2017].

Amnesty International. (2017). *Dangerously disproportionate: The ever-expanding National Security state in Europe*. Available from https://doc.es.amnesty.org/cgi-bin/ai/BRSCGI.exe?CMD=VERDOC&BASE=SIAI&DOCR=1&SORT=-FPUB&RNG=10&SEPARADOR=&&INAI=EUR01534217#2 [accessed 27 November 2017].

Becker, H. (1963). *Outsiders*. New York: Free Press.

Benedicto, J. and Morán, M. (2015). La construcción de los imaginarios colectivos sobre jóvenes, participación y política en España. *Revista de Estudios de Juventud*, 110(December): 83–104.

Bessant, J. (2017). Digital humour, gag laws, and the liberal security state. In: R. Luppicini and R. Baarda (eds), *Digital media integration for participatory democracy*. Hershey, PA: IGI Global, 204–221.

Blay Gil, E. (2013). El control policial de las protestas en España [Policing Protest in Spain]. *InDret: Revista para el Análisis del Derecho*, (4): 2–32.

Bondía, D. (ed.). (2015). *Defender a quien defiende: Leyes mordaza y criminalización de la protesta en el estado español*. Barcelona: Icaria

Cancio Meliá, M. (2014). El derecho penal antiterrorista español y la armonización penal de la Unión Europea. *Revista Justiça e Sistema Criminal*, 6(10): 45–72.

Camps, C. and Vergés, N. (2015). De la superación del miedo a protestar al miedo como estrategia represiva del 15M. *Athenea Digital*, 15(4): 129–154.

Chasco, S. (2014). Videovigilancia en el centro de Madrid. ¿Hacia el panóptico electrónico? *Teknokultura*, 11(2): 301–327.

CIS, Centro de Investigaciones Sociológicas. (1998). Survey 2315/1999. *Seguridad Ciudadana y Victimización (II)*. Madrid: Centro de Investigaciones Sociológicas. Centro de Investigaciones Sociológicas. [Online.] Available from www.cis.es/cis/opencm/ES/1_encuestas/index.jsp [accessed 10 July 2017].

CIS, Centro de Investigaciones Sociológicas. (1998). Survey 2284/1998. *Seguridad y Victimización (I)*. Madrid: Centro de Investigaciones Sociológicas. [Online.] Available from www.cis.es/cis/opencm/ES/1_encuestas/index.jsp [accessed 10 July 2017].

CIS, Centro de Investigaciones Sociológicas. (2007). Survey 2702/2007. *Delincuencia y victimización en la Comunidad de Madrid*. Madrid: Centro de Investigaciones Sociológicas. [Online.] Available from www.cis.es/cis/opencm/ES/1_encuestas/index.jsp [accessed 10 July 2017].

Cram, I. (2012). The war on terror on campus: some free speech issues around anti radicalization law and policy in the United Kingdom. *Journal for the Study of Radicalism*, 6(1): 1–34.

della Porta, D. (2015). *Social movements in times of austerity: bringing capitalism back into protest analysis*. London: Polity Press.

Departamento de Seguridad Nacional (DSN): *Informes seguridad nacional 2013, 2014, 2015*. Available from www.dsn.gob.es/es/estrategias-publicaciones/informe-anual-seguridad-nacional [accessed 27 November 2017].

Enriquez, E. and Centeno, M. (2012). State capacity: utilization, durability, and the role of wealth vs. history. *International and Multidisciplinary Journal of Social Sciences*, 1(2): 130–162.

Espinar-Ruiz, E. and González Río, M. (2015). Uso de internet y prácticas políticas de los jóvenes españoles. *Convergencias. Revista de Ciencias Sociales*, 69(September–December): 13–38.

Fassin, D. (2013). *Enforcing order: An ethnography of urban policing*. Malden, MA: Polity Press.

Feixa, C. and Nofre, J. (eds). (2013). *Generación indignada: topías y utopías del 15M [Indignant generation: topias and utopias of 15M]*. Lleida: Milenio.

Fernández de Mosteyrín, L. (2012). *La guerra contra el terror y la transformación de umbrales de violencia tolerada: Un estudio de la violencia en el País Vasco 1998–2010*. Universidad Complutense de Madrid. Available from UCM-Eprints: http://eprints.ucm.es/20050/.

Fernández de Mosteyrín, L. (2013). Rodea el Congreso: Un caso para explorar las bases del Estado Securitario. *Anuari del Conflicte Social 2012*, (2): 1129–1152.

Fernández de Mosteyrín, L. (2013). Las demandas de seguridad y las políticas antiterroristas tras el 11-S. In: M. Morán (ed.), *Actores y demandas en España: análisis de un inicio de siglo convulso*. Madrid. La Catarata.

Fernández de Mosteyrín, L. and Limón, P. (2017). [Forthcoming.] Paradigmas y políticas de seguridad: Una aproximación al Plan Estratégico Nacional de Lucha contra la radicalización violenta – PEN-LCRV 2015. *Política y Sociedad*, 54(3) (in press).

Flesher-Fominaya, C. (2011). The Madrid bombings and popular protest: misinformation, counter-information, mobilisation and elections after '11-M'. *Contemporary Social Science*, 6(3): 289–307.

Flesher-Fominaya, C. (2015). Debunking spontaneity: Spain's 15M/Indignados as autonomous movement. *Social Movement Studies*, 14(2): 142–163.

Flesher-Fominaya, C. and Cox, L. (2013). *Understanding European movements: new social movements, global justice struggles, anti-austerity protest*. London: Routledge.

Flesher-Fominaya, C. and Montañés Jiménez, A. (2014). Transnational diffusion across time: The adoption of the Argentinian Dirty War 'escrache' in the context of Spain's housing crisis. In: D. della Porta and A. Mattoni (eds), *Spreading protest*. Colchester: ECPR Press, 19–42.

Galdon-Clavell, G. (2015). Si la videovigilancia es la respuesta ¿cuál era la pregunta? Cámaras, seguridad y políticas urbanas. *EURE (Santiago)*, 41(123): 81–101.

García, S. and Ávila, D. (eds). (2015). Enclave de riesgo: gobierno neoliberal, desigualdad y control social. Madrid. Observatorio Metropolitano. Editorial Traficantes de Sueños.

García Galera, M. (2014). Jóvenes comprometidos en la red: El papel de las redes sociales en la participación social activa. *Comunicar*, 22(43): 35–43.

Garland, D. (2001). *The culture of control*. Oxford: Oxford University Press.

Gillham, P. (2011). Securitizing America: strategic incapacitation and the policing of protest since the 11 September 2001 terrorist attacks. *Sociology Compass*, 5(7): 636–652.

Goffman, A. (2014). *On the run: fugitive life in an American city*. Chicago: University of Chicago Press.

Hallsworth, S. and Lea, J. (2011). Reconstructing Leviathan: emerging contours of the security state. *Theoretical Criminology*, 15(2): 141–157.

Hickman, M., Thomas, L., Nickels, H. and Silvestri, S. (2012). Social cohesion and the notion of 'suspect communities': A study of the experiences and impacts of being 'suspect' for Irish communities and Muslim communities in Britain. *Critical Studies on Terrorism*, 5(1): 89–106.

Hueso, A., Boni, A. and Belda-Miquel, S. (2015). Perspectivas y políticas sobre la juventud en desventaja en España: Un análisis desde el enfoque de capacidades. *REIS Revista española de investigaciones sociológicas*, 152(October–December): 47–64.

Jackson, R. (2005). *Writing the war on terrorism: Language, politics and counter-terrorism*. Manchester: Manchester University Press.

Jarvis, M. and Lister, M. (2013). Disconnection and resistance: anti-terrorism and citizenship in UK. *Citizenship Studies*, 17(6–7): 756–769.

Jerez, A., Maceiras, S. and Maestu, E. (2015). Esferas públicas, crisis política e internet: El surgimiento electoral de Podemos. *História, Ciências, Saúde – Manguinhos*, Rio de Janeiro, v.22(December): 1573–1596.

Kundnani, A. (2015). *The Muslims are coming! Islamophobia, extremism and the domestic war on terror*. London: Verso Books.

Limón López, P. (2012). Producción jurídica e imaginación global: cartografías urbanas a través de la ley en Barcelona. *Geopolítica(s). Revista de estudios sobre espacio y poder*, 3(1), Monográfico: políticas urbanas en España: 117–135.

Limón López, P. (2013). Del No a la guerra a los Indignados: Los debates sobre el espacio público en las movilizaciones en la calle. In: M. Calvo-Sotelo and M. Luz (eds), *Actores y demandas en España: análisis de un inicio de siglo convulso*. Madrid: Los libros de la Catarata, 185–208.

Limón López, P. (2014). Imaginación geográfica y agencia política: produciendo espacio público a través del Derecho en Madrid (1992–2012). *EURE: revista latinoamericana de estudios urbano regionales*, 120: 183–200.

Llobert-Anglí, M. and Masferrer, A. (2015). Counterterrorism, emergency and national laws. In: G. Lennon and C. Walker (eds), *Routledge handbook of law and terrorism*. Abingdon: Routledge.

Lobera, J. (2015). De movimientos a partidos. La cristalización electoral de la protesta. *Revista Española de Sociología*, 24: 97–105.

Mann, M. (1984). The autonomous power of the state: its origins, mechanisms and results. *European Journal of Sociology*, 25(02): 185–213.

Ministry of the Interior. (2013, 2014, 2015). *Balance of criminality*. Madrid: Ministerio del Interior. Available from www.interior.gob.es/es/web/archivos-y-documentacion/archivos-y-documentacion [accessed 10 July 2017].

Mullard, M. (2007). Citizenship, globalisation and the politics of the war on terror. In: M. Mullard and B. Cole (eds), *Globalisation, citizenship and the war on terror*. (81–96.). Cheltenham: Edward Elgar Publishing.

Navarro Bonilla, D. and Esteban Navarro, M. (eds). (2007). *Terrorismo global, gestión de información y servicios de inteligencia*. Madrid: Plaza y Valdés.

Neocleous, M. (2000). *Fabrication of social order: A critical theory of police power*. London: Pluto Press.

Neocleous, M. (2011). 'A brighter and nicer new life': security as pacification. *Social & Legal Studies*, 20(2): 191–208.

Oliver Olmo, P. and Urda Lozano, J. (2015). Bureau-repression: administrative sanction and social control in modern Spain. *Oñati Socio-Legal Series* [online], 5(5): 1309–1328. Available from https://papers.ssrn.com/sol3/papers.cfm?abstract_id=2574670 [accessed 11 July 2017].

Pisarello, G. and Asens, J. (2014). *La bestia sin bozal: en defensa del derecho a la protesta*. Madrid: Los Libros de la Catarata.

Reinares, F. (2008). Tras el 11 de marzo: estructuras de seguridad interior y prevención del terrorismo global en España. In: C. Powell and F. Reinares (eds), *Las democracias occidentales frente al terrorismo global*. Madrid: Arial y Real Instituto Elcano de Estudios Internacionales y Estratégicos, 103–144.

Ríos, P., Rusiñol, P. and Marcos, P. (2001). Guillem Vidal alerta del peligro de infiltrar policía en las manifestaciones. *Diario El País* [online], 29 June. https://elpais.com/diario/2001/06/29/catalunya/993776847_850215.html [accessed 11 July 2017].

Rodríguez-Pina, G. (2012). Policías infiltrados en el 25-S: La Jefatura reconoce que había 'secretas' pero niega que provocasen la violencia. *Huffington Post* [online], 26 September. Available from www.huffingtonpost.es/2012/09/26/policias-infiltrados-en-e_n_1915348.html [accessed 11 July 2017].

Romanos, E. (2016). From Tahir to Puerta Del Sol to Wall Street: The transnational diffusion of social movements in comparative perspective. *Revista Española de Investigaciones Sociológicas*, 154(1): 103–133.

Soifer, H. (2008). State infrastructural power: approaches to conceptualization and measurement. *Studies in Comparative International Development*, 43(3–4): 231.

Subirats, J. (2015). Todo se mueve. Acción colectiva, acción conectiva. Movimientos, partidos e instituciones. *Revista Española de Sociología*, 24: 123–131.

Tarrow, S. (2015). *War, states, and contention: A comparative historical study*. Cornell: Cornell University Press.

Ulloa Rubio, I. (2012). El desafío de garantizar la seguridad pública. *Seguridad y Ciudadanía, Revista del Ministerio del Interior*, 7–8(January–December): 13–24.

Wacquant, L. (2009). *Punishing the poor: The neoliberal government of social insecurity*. Durham; London: Duke University Press.

Chapter 5

'Proxy repression'?

The causes behind the change of protest control repertoire by the Université du Québec à Montréal during the 2015 student strike

Luc Chicoine[1]

Introduction

The 2012 Québec student strike in protest against university tuition-fee hikes morphed into an expansive political action (Dufour 2012; Dufour and Savoie 2014). It transformed into a major political crisis for the incumbent government and led to an early election in which the Jean Charest Liberal government was defeated. In comparison, the later 2015 student strike against the austerity policies of the new Liberal government did not achieve anything similar in terms of its political impact. In order to explain this relative decrease in mobilisation, both in the number of striking students and in the mediatic visibility of the protests, it is necessary to consider the strategies deployed by public authorities in response to this mobilisation.

In this chapter, I demonstrate how the Charest government transferred a good deal of the responsibility for managing student protests over to university managers. Here, I examine the case of Université du Québec à Montréal (UQAM) – an institution with a reputation both for student activism and for its historical approach of adopting a 'softly softly' approach to student protests. This posture was to change dramatically in 2015. The spectrum of new aggressive tactics used by the university to repress students was wide ranging: omnipresent on campus video surveillance; intimidating private security guards (sometimes dressed in civilian clothes); injunctions; social media monitoring; pre-emptive disciplinary measures against student leaders; mass arrests. All were used by UQAM's management to control the protests.

How can we explain the development of this new approach by universities to managing student protests (Gillham *et al.* 2012)? Why does a university decide to adopt repressive[2] measures against its students who engage in legitimate political action? Repressive responses by educational institutions are best explained by the 'logic of path dependence' (Pierson 2004; Ancelovici 2013). By this I mean that provincial authorities learned lessons from what they did during the 2012 student strike and this influenced their decision to adjust their repressive tactics in 2015. Thus, the implementation of new protest control mechanisms by Université du Québec à Montréal administrators can be considered an extension of the repressive model used by the Québec government during the 2012 strike.

Three different research approaches were used to carry out this study. As a member of the Academic Senate from 2012 to 2016, a member of the Council of the Faculty of Humanities, and a member of the disciplinary committee created in the wake of the 2015 strike, I had a privileged insight that allowed me to gain in-depth knowledge of the institutional dynamics at play. This also allowed me to gain 'insider' observations of certain key episodes.

Those observations were enhanced and confirmed by three exploratory semi-structured interviews with two students (Antoine and Alice) and a university teacher (Zinedine) – who also sat on the board of directors – between 23 and 28 February 2017, in Montréal. They were chosen for this study because of their intimate knowledge of the protest and/or governance dynamics at UQAM and because they had been witnesses to or participants in some of the numerous 'class interruptions'[3] during the 2012 and 2015 strikes, thus constituting the basis of a purposive snowball nonprobabilistic sample.

Finally, I used the data of a protest event analysis (PEA) linked to higher education in Québec between 1 January 2005 and 31 December 2016.[4] Using the keywords *éducation, étudiants, enseignants, manifestation, sit-in, bed-in, protestation, campement, pétition, marche, occupation, grève, blocage, rassemblement, squat, caravane*, I researched articles from the Montréal daily newspaper *La Presse* and drafted a list of 633 such events. I then coded the locations and targets of the strikers based on variables relevant to this study.

Using this data, in this chapter, I first examine the state's repressive tactics used during the 2012 unlimited strike. Three main strategic lines of governmental action were developed to counter student mobilisation: delegitimising the student movement, repressing protest actions and criminalising student dissent. Then, I outline the evolution of these tactics for the 2015 students' protests, particularly in regard with the reaction of UQAM's senior management. Finally, drawing on the literature on repression and social movements, I show how the state benefited from these changes.

Repertoire of government counter-strategies during the 2012 strike

The Liberal government used many different tactics to repress the striking students in 2012. The following short review of these manoeuvres serves to provide the background to understand the changes that occurred in the next unlimited strike.

The delegitimisation of the student movement

Every student strike in Québec was characterised by a fierce battle to influence public opinion. During the last series of student strikes (2005, 2007, 2012 and

2015), one of the first strategies used by the government was to represent the students as violent bullies. These public condemnations functioned as a 'threat amplification' (Monagham and Walby 2012), a framing process whereby a protest group were described as criminals. It was a rhetorical strategy that empowered the state by 'justifying' their recourse to repressive – and sometimes pre-emptive – measures against students.

This went well beyond discursive tactics; it involved the government setting a certain ultimatum. The strategy was simple: the government asked student associations to condemn their members' violent actions, otherwise the government would not negotiate with them.

The government hoped to divide and rule by encouraging some national student associations to accept this official demand while knowing others would not, thus dividing students and thereby reducing their leverage, as was the case in 2005. This move also worked to divide protesters in the streets, with some opposing others about the place of 'acts of vandalism' in protest action. Finally, and equally importantly, governmental insistence that student actions represented unacceptable criminal violence also functioned to discredit the movement. It was a rhetorical device that positioned the government as guardian of law and order, safeguarding the public against hordes of dangerous rioters. In this way, anybody wearing the red square[5] was associated with violence (Nadeau 2012).

Another governmental strategy was to deny the students' right to strike. Overnight, the government stopped using the word 'strike' and started using the word 'boycott'. This semantic shift diminished the political value and importance of the uprising against the government by comparing their actions with difficult and ungrateful customers who were refusing a service. In a society where everyone has to contribute their 'fair share',[6] only 'spoiled brats' would act as they had. It was a theme that was mimicked by a large number of media columnists and pundits.

All these comments, opinions and attitudes combined to form a dominant discourse that substantially increased the difficulty of mobilising students and allies. Moreover, this *soft repression* – including stigmatisation, ridicule and attempts to silence students (Ferree 2004) – set up an already polarised basis of discussion for the next mobilisation; an effect clearly observable in the 2015 strike.

The judiciarisation of the conflict

In 2012, new legal means were adopted to repress student protest actions. (Lemonde *et al.* 2014; Dufour 2016). Up to that point, the student association's right to strike had neither been admitted nor forbidden by law, given that it was perhaps inevitable that those opposed to the student action would use legal mechanisms to discredit the campaign and make strike action very difficult.

A first step towards this involved those students who opposed the strike asking Québec's Superior Court for an injunction that would allow them to attend their classes despite the blockages and the class interruptions. Close to 50 such injunctions forcing specific courses to be held were granted by the Superior Court over the course of the strike. These injunctions placed education institutions in a delicate position since they then were responsible for providing classes to the individual students who requested it. Several universities and colleges had to call the police to remove strikers blocking access to classrooms or buildings, which at times escalated into violent clashes, notably at Université du Québec en Outaouais where 151 students were arrested on a single day.

On 18 May 2012, the Charest government adopted Bill 78, named 'An Act to enable students to receive instruction from the post-secondary institutions they attend'. It imposed legal constraints across the board: students were prohibited from 'impeding' their education or that of their peers, while teachers and their unions were prohibited from participating in a concerted action that would 'impede education'. Higher education institutions were now obliged to suspend the winter and summer terms and resume classes no later than 30 August. The entire population of Québec was also obliged to warn the police of any protest or gathering of more than 50 people, in clear violation of their right to free expression. This 'bludgeon law' also made provisions for individual fines of up to CAD 1,000 and up to CAD 125,000 for organisations (student associations, schools, unions, etc.).

The law was strongly condemned by the Commission des droits de la personne et des droits de la jeunesse, Amnesty International, the United Nations' High Commissioners of Human Rights, central labour congresses and many more civil society organisations. They all considered it a serious infringement of fundamental liberties guaranteed by the Québec Charter of Human Rights and Freedoms. Large segments of the population also condemned the law with citizens mobilising spontaneously against it. They held noisy 'pots and pans' protests in residential neighbourhoods every evening, imitating the Chilean *cacelorazo* for a few weeks (Drapeau-Bisson *et al.* 2014).

Another way the student political action was criminalised was through the implementation of municipal bylaws aimed at managing protests. Montréal adopted bylaw P-6, prohibiting the wearing of masks and forcing organisers to provide the itinerary of their marches at least 24 hours in advance, otherwise their protests would be deemed illegal and all participants would be punishable by a fine between CAD 500 and CAD 3,000.

Repression of protest actions

In 2011, four student militants, three of whom were members of an important national student association's executive, were arrested. It was revealed that the arrest was the result of an operation by Montréal police's GAMMA unit ('*Guet des activités et des mouvements marginaux et anarchistes*' [Surveillance of marginals and anarchists activities and movements]), that targeted anarchist and left-leaning groups. Internal documents obtained through the Freedom of Information Act have since revealed that, according to that police unit, 'the student body is a recruitment pool for such groups' and that they 'have ties with associations at CEGEP du Vieux-Montréal and UQAM' (Dominique-Legault 2016). Although this unit was soon disbanded once its existence had become public, the subsequent actions taken by the Montréal police suggest that Montréal's student associations had been under surveillance.

Québec's provincial police, the Sûreté du Québec, had a similar unit, the *Division des enquêtes sur la menace extrémiste* (DEME) [Extremist threat investigation division that investigated student leaders]. This representation of students-as-threats by police services specialised in criminal intelligence begins to explain why we saw an explosion in the number of arrests, which, in 2012, exceeded 3,500. This figure was an increase of approximately 20 times the number of arrests made during the previous strike.

One police tactic used during the Maple Spring that also explains such astonishing increases in arrests was the practice of encirclement – also known as 'kettling', containment or corralling. During protests, police officers encircled large groups of people, sometimes several hundred individuals. Those who were trapped had to wait on-site for several hours, often late into the night, handcuffed, until they were given a hefty fine related to a municipal bylaw such as P-6, or to the Highway Safety Code. About 30 such major operations were identified during this conflict (Ménard *et al.* 2014: 17).

A similar policing method involved arresting students who were identified as 'threats' – through political profiling and surveillance of previous protest actions – before they made their way to a protest. Aside from this use of 'preventive arrests', many individuals wearing a sign that could identify them as supporters of the students' cause, notably the red square, were searched or arrested without just cause.

The use of crowd control weapons – pepper spray, sound bombs, batons, CS gas and others – were widely used, as documented by the Special Commission to Examine the Events of Spring 2012 mandated by the government a few months after the events (Ibid. 2014). Even if no deaths were reported, there is evidence of serious physical injuries inflicted on the protesters (two protesters each lost an eye in two separate events) (Ibid. 2014).

The government paid a steep price for those repressive operations. According to the report of the Special Commission to Examine the Events of Spring 2012 mandated by the Public Safety Minister on the student protest campaign of 2012, the total expenses incurred by police in attempting to manage the conflict totalled nearly CAD 26 million for the province (Ibid. 2014: 105).

The end of tolerance by higher education senior managers

Bill 78 and the interlocutory injunctions both put enormous pressure on higher education executives and teaching staff, who were legally obliged to teach classes in quite abnormal conditions. Striking students refused to comply with these legal obligations and blocked access to courses, forcing senior managers in these institutions to either disobey the law or call the police to dislodge them. For example, Université de Montréal blindly applied Bill 78, which led to two days of violent clashes between the 'riot control unit' and students in the university's halls. In the end, these governmental pressures on universities did not prove to be very effective in constraining the strikes and protests. Yet, the government was not about to abandon this policy of repression as it offered many advantages, as I demonstrate in what follows.

The 2015 strike: video surveillance and intra-institutional repression

The protest 'management' measures implemented by UQAM's authorities in 2015 described below best fit della Porta and Diani's description of 'hard' responses in their typology (2006: 198). We see here evidence of escalated conflict and the use of force rather than any attempt to use a 'preventive' or 'de-escalation' approach (McPhail *et al.* 1998). In the following section, I will demonstrate how this shift away from a 'negotiated management approach' (Vitale 2005) manifested itself in a university deemed open to dissent.

Class interruption at UQAM

In 2012 and during previous student strikes since the 1960s, interruptions to classes were generally not problematic. As Alice explained: 'Class interruptions went well in 2012. […] At some point, they were a matter of fact, they almost became symbolic.' This seems largely to have been due to the institution's approach in dealing with student protesters. Another interviewee put it this way: 'Up to now, in my 33 years at the institution, there were strikes, class interruptions, but everything was negotiated and things rapidly went back to normal. There weren't, there never was any violence, etc.' (Zinedine). Yet, things seemed different in 2015; by then, as Zinedine explained: '… it felt like there were orders emanating directly from the office of the prime minister in Québec. […] At that moment, we felt a real will to break the movement, we can talk about true repression.'

Expulsion threats

On 20 March 2015, nine UQAM students received notice from the university administrators that they risked a lifetime suspension following what was described as 'vandalism and illegal acts' committed on campus as far back as 2013. All of the students were involved in university life, including one who sat on the university board of directors. On the same day as the notice was issued, another student member of the board lost his seat after learning he was no longer a student after the office of the vice-rector used its powers in an exceptional manner by denying that student his request for a registration extension for his Masters' programme. It's not yet known whether this was a move to stack the board or to silence dissenting students, but it happened at a time in which this governing instance had a specific role to play in the repression of striking students, as we'll discuss later.

The shock was brutal for UQAM students, as those procedures came into force only three days before the beginning of a strike that several student associations had voted in favour of starting. These measures not only had a 'chilling effect' (Schauer 1978) on other students, they also destabilised the leadership of the strike. On 1 April, during a radio interview, the education minister declared: 'I've said the following to all rectors: "You have the means to act, implement measures, impose sanctions, even if it's only two or three students who take things way too far, who exaggerate, on a daily basis"' (transcribed from Robillard's 2015 story published in *La Presse* on 1 April).

Video surveillance

Evidence of 'student disruption' was accessed mainly from footage from surveillance cameras installed all over the campus. They had been installed in the wake of the Maple Spring as part of the university's efforts to damp down dissent. Student associations – and various unions in the university – demanded better guarantees than those offered by the university regarding the use of the images recorded by those cameras. UQAM's management unsuccessfully tried to reassure the university community about the legitimacy of its conduct while staunchly sticking to its policy. The nine disciplinary hearings confirmed the associations' fears about the university's true intentions. This attack on student strikers directly shaped the rest of the conflict at UQAM.

Blocking of central pavilions and injunction

On 30 March 2015, in retaliation against these disciplinary hearings, students conducted a power grab and blocked, for several hours, all access to the university's main buildings. The university immediately countered by asking the Superior Court for an interlocutory injunction that prohibited any disruption of its activities under penalty of obstructing justice, which can result in a sentence of up to a year in jail. This particular injunction legally restricted the strikers' repertoire of actions by making class disruptions illegal. Clearly stung, UQAM students abandoned, almost overnight, their anti-Austerity stance and refocused their struggle against the university's repression and demanded the abandonment of the disciplinary hearings against the nine students.

Delegitimisation: masked 'commandos' at UQAM?

Despite the injunctions and cameras, striking students continued to interrupt classes. To remain unidentifiable and avoid the serious punishments attached to non-compliance with the injunction, they dressed in black and covered their faces. As Antoine explained:

> The state of repression at the university was such that it was dangerous to do class interruptions without a mask, so to protect ourselves, it was important [...] to wear a mask. People didn't do that lightly. It has consequences. It provoked a backlash from people who don't fully understand the nature of this action [...] and it gives a false impression, you don't feel you're doing something wrong by wearing a mask, but that's the message people receive.

This new need for disguise demobilised some protesters. As Antoine observed: 'Some students wanted to participate, they understood why people were wearing masks, but they weren't ready to do so themselves. It's psychologically hard to wear a mask.'

Wearing masks also compounded the perception that those people who were interrupting the classes were violent, intimidating individuals who had no respect for the inherent values of the university. Yet, this move to create fear was part of a larger process designed to delegitimise student protests that began before the strike itself had started, notably with the publication of a collective letter signed by 14 political science teachers in a Montréal daily newspaper. Entitled 'À l'UQAM, l'intimidation doit cesser!' [Intimidation must stop at UQAM!], the teachers' letter used the same delegitimisation themes being used by the Charest government in 2012. The group letter claimed that:

> Over the past few years [...] our university has been the victim of the acts of a minority: class interruptions by sometimes masked self-proclaimed commandos, intimidation, harassment, scuffles, vandalism, ransacking, disruption of meetings and conferences, repeated strikes. Such actions threaten university life itself as well as the UQAM's particular character.
>
> (*Le Devoir*, 24 February 2015)

The initiative came from teaching staff but, coincidentally, in December 2014, I spoke with a vice-rector to ask him to support a resolution I was to present to the Academic Senate concerning sexual harassment policy. To my great surprise, this executive, in charge of the university's Security and Prevention Service (SPS), tried to negotiate his support for the resolution in exchange for my public denunciation of the 'violent and intimidating' behaviour of students on campus. Obviously, I refused the deal. In the end, he publicly supported the teachers' group letter, despite the harm it could cause to the university's reputation (*Le Devoir*, 25 February 2015).

That letter, signed by political science teachers, seems to have greatly influenced negative media coverage of the strike, with some journalists reusing the term 'commando' in reference to the masked students. This negative press can partly explain that spring's forceful repression as several studies have demonstrated there is an increase of repression when a protest movement is not widely covered by the media or is covered in a highly unfavourable light (Wisler and Giugni 1999; Koopmans 2005).

Indeed, from day one of the strike, the university's SPS drastically changed its approach in managing class interruptions. Instead of letting the interrupters go freely from class to class, it tried to stop them. New, intimidating private security guards were brought to the school. As Antoine explained: 'They hired specialised "strike busting" agencies […] they sent goons, people who, I am not kidding, weighed 300 pounds.' These agents were generally posted in front of classroom doors in order to block groups wishing to enter for a class interruption: '[In 2015] UQAM decided to block the doors before the student protesters arrived, but in other universities, it's the students [who are involved in the protests] who block the doors!' (Antoine). The security agents on campus often tried to physically unmask the protesters so they could be filmed and their identity revealed. Other security guards were dressed in civilian clothes and sat in classrooms, filming events.

Mass arrests

These tactics inevitably led to an escalation of confrontations between students and security guards employed by the university. This culminated on 8 April 2015, when university officials called the police to arrest a group of protesting students who were interrupting classes. Violent scuffles occurred between security agents and students and 22 of them were arrested and charged with misdemeanours or unlawful assembly.

In the media that day, the university rector justified his decision to call the police, saying they were reacting to 'attacks' and 'threats' on the part of students. During the subsequent trial of the 22 accused, it came to light that UQAM were given a choice by the Montréal police: either 'forcibly evicting all intruding occupants' without arresting them, or 'arrest those individuals' on campus. The decision to arrest was justified by the university administration as a desire to 'find out who the masked individuals were', as the director of the SPS testified under oath (audio recording of court proceedings).

Once police removed the arrested students, a large number of other students and teachers gathered in the pavilion. In the meantime, Rector Proulx left the university to be interviewed on the radio about the events that had occurred. He also had to respond to the prime minister's comments that: 'We remind university managers to use security forces if necessary. It's their duty' (Ici Radio-Canada Première, 8 April 2015).

Upon his return to the university, the rector refused to address the students. Thus, the gathering of students and teachers spontaneously morphed into an occupation of the pavilion.

Late that evening, some occupants expressed their frustration and anger by damaging property which led to police mobilisation around the university. As Zinedine, who was there at the time, explained: 'I told the students: "Watch out! At five past midnight, the police is going to intervene". Why? Because one of the vice-rectors has assured us the police would not intervene again that day.' This prediction came true just as several students were calmly leaving the building:

> All the doors are open on each side. Students are calmly leaving the building. All of a sudden, the police arrived, and I saw some officers destroy the pavilion's windows to enter the building, and that's what they [the TV crews] filmed, but in fact, there was no one inside. [...] It was completely staged! That's when I started to understand that, watch out, the reason why the rector had refused [to speak with the protesters] was because he actually had directives that came straight from the government.
>
> (Zinedine)

Québec's involvement: 'proxy repression'?

In this section, I analyse the influence of the provincial government on UQAM's senior management choice of repressive repertoire. I call this influence 'proxy repression' to describe the state's strategy to pressure a group or organisation so that it applies repressive measures against dissenters.

Analysis of UQAM's vulnerabilities

How can we explain why the UQAM senior administrators – a university with a reputation for being open to protests and, generally, an ally of social movements – decided to engage in increasingly repressive tactics? Currently, there is little literature on this subject. According to Walker *et al.* (2008), universities are vulnerable to nonparticipation and delegitimisation strategies, which leads them to use conflict to counter them. Since UQAM's reaction to the protesters hardened considerably between 2012 and 2015, we must conclude that UQAM's vulnerabilities to student protests greatly increased during that period. However, Walker didn't outline the nature of the vulnerabilities in question. So, in order to make sense of the variation here, we still need to determine which vulnerabilities changed in this specific time period (2012–2015).

Lorenzo Cini (2016) hypothesised that the repression of student mobilisations is stronger when a university's management is more 'managerial' than 'academic'. This explanation, however, does not appear to fit with what happened at UQAM. First, there was no significant variation in the structure of the university's management between 2012 and 2015 that could explain the change in the level of repressive actions. Second, the personality, background and opinions of the university rector does not seem to fit the hard-line managerial type.

The criminalisation of student strikes and the use of injunctions forcing universities to dispense classes did put pressure on universities' senior managers to deal harshly with student dissent. Yet this does not fully explain UQAM's reaction in 2015 because we know it requested the injunction from the Superior Court, and other universities and colleges affected by this strike did not employ such measures.

It also seems that UQAM's financial vulnerability provides another explanation for the changes it made to managing student protest. My interviewee Zinedine, a board member, was clear that: 'Nowadays, all that matters are financial issues. It's cruel and terrible, I find it very depressing.' When asked if it was the board that made the decisions during the crisis, he answered without hesitation: 'No. It was Québec. It's a little strange to say it that way, it seems like a sweeping statement. What I mean is that there was so much pressure from the government that we only did what we could.'

Aside from almost a decade of university funding cuts, one of the great financial vulnerabilities of UQAM lies in an agreement between the Ministry of Education and UQAM following certain catastrophic real estate developments. The agreement imposed a hefty financial penalty (involving the removal of a conditional grant totalling CAD 25 million) on UQAM if the institution did not rebalance its budget. In 2015, the establishment's management was aware that this would not be achieved in the 2015–2016 fiscal year and had to obtain the minister's approval for its 2016–2019 Deficit Resorption Plan.

The intervention of UQAM's president of the board, Lise Bissonnette, on TV a few days after the 9 April events suggests a link between the government's financing of universities and their approach to managing protest activities. Bissonnette stated that, even though the education minister said 'on-campus safety' was the rectors' responsibility, the government had little respect for that autonomy when making budgetary cuts 'and things like that' without consultation (Ici RDI, 13 April 2015).

The fact the university was in a weak financial situation in 2015 meant the government could exercise considerable leverage by requiring UQAM's management to comply with what it required. In short, had the university refused, the education minister had the option of withholding a conditional grant funding or even putting the university in trusteeship, arguing it was being badly managed.

The advantages of proxy repression

Why did the government insist that universities control student strikes? Although it is impossible to definitively answer that question, social movement theory points to a few hypotheses.

First, 'proxy repression' tends to shift the target of the protest action to the institution which implements the repressive measures. It is an oft-observed effect of repression that it leads to protest actions aimed at the repressive measures themselves, as was the case with the pots and pans demonstrations in 2012 (Drapeau-Bisson *et al.* 2014). Figure 5.1 demonstrates the effect of repression on the evolution in the targeting of protest actions. During the 2005 strike, legal injunctions had not been sought or issued by the courts; the senior managers of education institutions were still exhibiting a high level of tolerance in the face of 'class interruptions'. This probably explains the absence of protest actions targeting education institutions. In 2012, however, there were at least 25 protest events and most are linked to university management deciding to enforce either a legal injunction or Bill 78. Senior managers at UQAM were targeted by 67 per cent of protest events linked to the 2015 strike: this clearly demonstrates how repressive measures can totally refocus a protest movement.

Since there is often a correlation between the targets of the protesters' actions and the places where those protest actions take place, it is possible that the evolution of the political process risks being geographically limited to campuses most involved, as Figure 5.2 illustrates clearly.

Although universities are open institutions, protests that occur on their campuses are generally not covered by the media as well as those that occur in public spaces (Earl 2003). The net effect of this shift in the locus of the protest's targets and their geographic confinement is to make them less visible in the media and this can lead to an increase in the scope and scale of repressive measures.

This visibility deficit has other benefits for repressive governments, the most important of which is that they don't have to pay the political price of using severe repressive measures – something which may not go down well in a democracy (Davenport 2007).

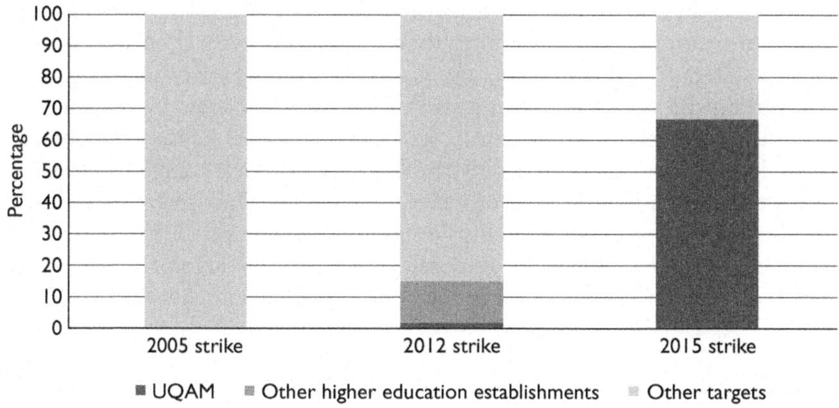

Figure 5.1 Targets of protest actions during student strikes.

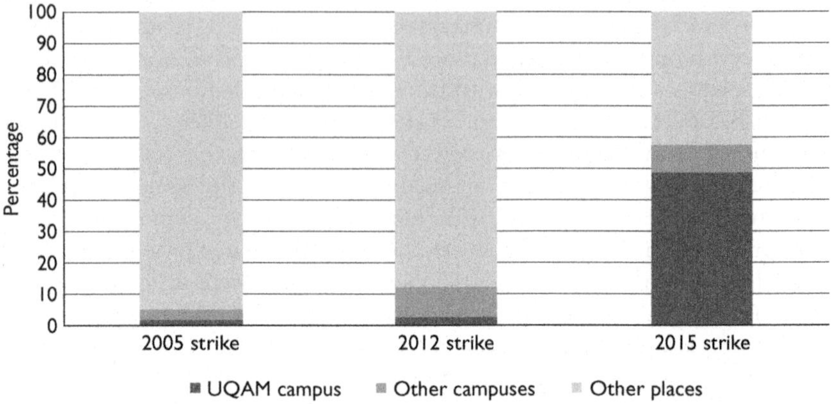

Figure 5.2 Localisation of protest actions during student strikes.

Conclusion

I argue here that we can explain how and why some universities become directly involved in repressing student political dissent by pointing to the vulnerabilities of education establishments. Moreover, it is highly likely that the economic vulnerabilities which university managers now need to deal with are themselves the consequences of governments imposing neo-liberal 'reforms' aimed at neo-liberalising universities. Demonstrating such an effect is outside my brief here and will only occur as a result of more analytically focused research. It seems too that these processes of neo-liberalising higher education have the effect of redefining students as 'client-investors', thereby denying legitimacy to any collective political actions they may take.

The 'subcontracting' of the surveillance and governance of student protests to universities by the state seems at first glance to transfer more responsibilities to these institutions, thus promoting their autonomy, even though it is quite unclear which part of the fundamental mission of a university it fulfils.

Subverting the autonomy of universities in this way occurs by reducing the independence and political freedoms of its various members – professors, students and employees – and affects not only the capacity of young people to engage in political expression, but also that of society as a whole. This is why resistance to these anti-political processes will only be successful when everyone in the university works together to rebuild a university conceived as an institution of the common dedicated to knowledge and democracy (Dardot and Laval 2014).

Notes

1 I would like to extend my most gracious thanks to Mare-Lise Drapeau-Bisson, Marcos Ancelovici, Caroline Turcotte-Brûlé, Pascale Dufour, Marion Sirieix, Charles

Carrier-Plante and to all other members of the *Chaire en sociologie des conflits sociaux*. Without your help, this chapter would never have been possible.

2 I will adopt Earl's definition of repression 'as state or private action meant to prevent, control, or constrain non-institutional, collective action (e.g., protest), including its initiation' (2011: 263).

3 In order to enforce the strike mandates, local student associations organise class interruptions. This means gathering small groups of students who then visit each classroom to let the attending teachers and students know that the strike has begun and to make noise until they leave. Teachers, observing that teaching conditions are not met, generally promptly leave their classroom, which cancels the class.

4 This PEA, I was involved in, was produced within the context of a wider research project titled 'The Institutional Foundations of Contentious Politics: A Comparative Analysis of Social Struggles in Québec, France, and Spain', SSHRSC Insight Grant, 2016–2021, led by Pascale Dufour and Marcos Ancelovici.

5 The red square was the symbol of the protest against the rise in university fees. Made of red cloth, it was often worn pinned to the chest.

6 This slogan was used by the finance minister when tabling his 2011 budget as a justification for an across-the-board cost increase for public services, including tuition fees.

References

Ancelovici, M. (2013). The origins and dynamics of organizational resilience: A comparative study of two French labor organisations. In: P. Hall and M. Lamont (eds), *Social Resilience in the Neoliberal Era*. New York: Cambridge University Press.

Cini, L. (2016). Student struggles and power relations in contemporary universities. The cases of Italy and England. In: R. Brooks, *Student politics and protest: international perspectives*. New York: Routledge.

Dardot, P. and Laval, C. (2014). *Commun. Essai sur la révolution au XXIe siècle*. Paris: La Découverte.

Davenport, C. (2007). State repression and political order. *Annual Review of Political Science* 10: 1–23.

della Porta, D. and Diani, M. (2006). *Social movements: An introduction*. Second Edition. Oxford: Blackwell.

Dominique-Legault, P. (2016). Des savoirs policiers sur les 'mouvements marginaux'. Les constructions du projet GAMMA du SPVM. *Criminologie*, 49(2): 301–321.

Drapeau-Bisson, M-L., Dupuis-Déri, F. and Ancelovici, M. (2014). 'La grève est étudiante, la lutte est populaire!' Manifestations de casseroles et assemblées de quartier. In: F. Dupuis-Déri and M. Ancelovici (eds), *Un printemps rouge et noir*. Montréal: Écosociété.

Dufour, P. (2012). Politique de la rue contre politique des urnes? Le mouvement étudiant québécois du printemps 2012 et la question de la représentation politique. *Savoir/Agir*, 22(4): 33–41.

Dufour, P. (2016). Mobilisation du droit dans le conflit étudiant de 2012 au Québec: quand le juridique se mêle de la contestation politique. In: D. Lamoureux and F. Dupuis-Déri, *Au nom de la sécurité. Criminalisation et pathologisation des marges*. Montréal: M. Éditeur.

Dufour, P. and Savoie, L-P. (2014). Quand les mouvements sociaux changent le politique. Le cas du mouvement étudiant de 2012 au Québec. *Canadian Journal of Political Science [Revue canadienne de science politique]*, 47(3): 475–502.

Earl, J. (2011). Political repression: iron fists, velvet gloves, and diffuse control. *Annual Review of Sociology*, 37: 261–284.

Ferree, M. (2004). Soft repression: ridicule, stigma, and silencing in gender based movements. In: D. Myers and D. Cress (eds), *Authority in Contention*, 25: 85–101.

Gillham, P., Edwards, B. and Noakes, J. (2012). Strategic incapacitation and the policing of Occupy Wall Street in New York City, 2011. *Policing and Society*, 23(1): 81–102.

Ici Radio-Canada Première, Le 15–18. (2015). 8 April, 15:00. Available from http://ici. radio-canada.ca/emissions/le_15_18/2013-2014/chronique.asp?idChronique=368783 [accessed 20 November 2017].

Ici RDI, 24/60. (2015). 13 April, 19:00. Available from http://ici.radio-canada.ca/nou velle/715651/bissonnette-lise-robert-proulx-recteur-uqam [accessed 20 November 2017].

Koopmans, R. (2005). Repression and the public sphere: discursive opportunities for repression against the extreme right in Germany in the 1990s. In: C. Davenport, H. Johnston and C. Mueller (eds), 2005. *Repression and Mobilization*. Minneapolis: University of Minnesota Press.

Lemonde, L., Bourbeau, A., Fortin, V., Joly É. and Poisson, J. (2014). La répression judiciaire et législative pendant la grève. In: F. Dupuis-Déri and M. Ancelovici (eds), *Un printemps rouge et noir*. Montréal: Écosociété.

McPhail, C., Schweingruber, D. and McCarty, J. (1998). Policing protest in the United States: from the 1960s to the 1990s. In: D. della Porta and H. Reiter (eds), *Policing Protest: The Control of Mass Demonstrations in Western Democracies*. Minneapolis: University of Minnesota Press.

Ménard, S., Grenier, B. and Carbonneau, C. (2014). *Rapport. Commission spéciale d'examen des événements du printemps 2012*. Québec: Publications du Québec.

Monagham, J. and Walby, K. (2012). Making up 'terror identities': security intelligence, Canada's Integrated Threat Assessment Centre and social movement suppression. *Policing and Society*, 22(2): 133–151.

Nadeau, J-F. (2012). 'Le carré rouge de Fred Pellerin: «violence et intimidation», affirme la ministre de la Culture'. *Le Devoir*, 9 June. Available from www.ledevoir.com/culture/ actualites-culturelles/352046/le-carre-rouge-de-fred-pellerin-violence-et-intimidation-affirme-la-ministre-de-la-culture [accessed 20 November 2017].

Pierson, P. (2004). *Politics in Time: history, institutions and social analysis*. Princeton: Princeton University Press.

Robillard, A. (2015). 'Expulsions d'élèves: Blais nuance ses propos'. *La Presse*, 1 April. Available from www.lapresse.ca/actualites/education/201504/01/01-4857561-expulsions-deleves-blais-nuance-ses-propos.php [accessed 20 November 2017].

Schauer, F. (1978). Fear, risk and the first amendment: unraveling the chilling effect. *Boston University Law Review*, 58: 685–732.

Vitale, A. (2005). From negotiated management to command and control: how the New York Police Department polices protests. *Policing and Society*, 15(3): 283–304.

Walker, E. T., Martin, A. and McCarthy, J. (2008). Confronting the state, the corporation, and the academy: The influence of institutional targets on social movement repertoires. *American Journal of Sociology*, 114(1): 35–76.

Wisler, D. and Giugni, M. (1999). Under the spotlight: the impact of media attention on protest policing. *Mobilization* 4(2): 171–187.

Chapter 6

Governing, monitoring and regulating youth protest in contemporary Britain

Sarah Pickard

Introduction

At the start of the new millennium, the Human Rights Act 1998 came into force in Britain. It incorporated into British law articles 10 and 11 of the European Convention on Human Rights (Council of Europe 1950: 5–6) that protect 'the right to freedom of expression' and 'the right to freedom of peaceful assembly'. Yet, as this chapter will show, these rights are regularly curtailed.

Young people have been the foremost protagonists in numerous political protests in the twenty-first century. During the 13 years of Labour power (under Tony Blair 1997–2007 and Gordon Brown 2007–2010), two major waves of protest took place. First, the Stop the War Coalition (StWC) protests against a British invasion of Iraq as part of the US-led coalition forces brought over a million demonstrators to London on 15 March 2003 (Bloom 2012), including many young people (Cunningham and Lavalette 2004). Second, the smaller G-20 London summit protests in April 2009 came shortly after the 2008 global banking and financial crises.

When the Conservative–Liberal Democrat coalition government was in power, from May 2010 to May 2015, there was an increasing amount of youth-led dissent centred on youth issues. Grievances included the rise in university tuition fees, cuts to the higher education budget, the abolition of the Education Maintenance Allowance (EMA), youth unemployment and a lack of affordable housing. More broadly, there was opposition to the austerity policies, public service cuts (including big reductions in spending on youth services), taxes favouring the political and business classes, as well as concerns about the failure to clamp down on tax avoidance and to address environmental degradation (Pickard 2015, 2018c).

A series of marches, occupations and other forms of direct action occurred in numerous cities in 2010 (Olcese and Saunders 2014), followed by smaller student-led days of action each November. In 2011, a large anti-public-sector cuts march took place and Occupy London camped outside Saint Paul's Cathedral through the winter of 2011–2012. Students were also engaged in a series of occupations and other direct actions on university premises against austerity measures affecting university staff, in late 2013 and 2014. Lastly, there were environmental protests involving environmental networks, such as the 'Frack-off' and 'No dash for gas' campaigns against shale extraction (Pickard 2018b).[1]

Youth-led protest took varied forms, involving different degrees of engagement (Grasso 2017) from (e)petition-signing, social media campaigns and non-violent direct action, such as marches and demonstrations to civil disobedience, including occupations (of public and private buildings and spaces), lock-ons, tripods and encampments as well as, to a much lesser degree, violent actions with attacks on buildings and police officers (Pickard 2018a).

In response to this ongoing political dissent on the part of young people, the British state enacted various measures. The government introduced new legislation and amended existing criminal justice law on protests. The state also increased surveillance and monitoring measures of protesters. Lastly, the police developed new methods of policing protests. These striking and intertwined evolutions in the regulation and governance of young people's political dissent raise crucial issues pertaining to the democratic right to protest and the potential consequences of dissent in an ostensibly liberal democracy that is Britain.

This chapter documents evolutions in the regulation of protest actions and protesters in contemporary Britain in which young people were the main protagonists. I first scrutinise changes in government legislation pertaining to public protests. Then, I examine developments in policing methods of protests and especially evolutions in the monitoring of protesters through the lens of youth-led protest in the twenty-first century.

Increasing the governance of protest within the legal framework

While the Human Rights Act 1998 marked an encouraging constitutional turning point in Britain for the protection of the right to protest, over the past decades, there has also been a growing amount of legislation both to restrict protest and to increase the surveillance of protest and protesters through data collection. This section outlines the main changes.

The Public Order Act 1986, introduced by the Conservative Government of Margaret Thatcher, created several new public order offences, including 'violent disorder' and established certain restrictive prerequisites for a peaceful protest to take place. Namely, at least six days prior to the event, organisers of a 'procession' (a march) must inform the local police of the date, time and route, as well as the name and address of at least one organiser. The Public Order Act 1986 (sections 12–14) allow the police to impose conditions on marches and demonstrations in a public place. The police authorities can change the location of a march, limit how long a march lasts, limit the number of people attending, stop a sit-down protest if it blocks a road or public walkway and set other conditions. The police also have the power to forbid a march by applying a banning order, or the dispersal of a public assembly if a senior police officer 'has reasonable belief' that it may result in serious public disorder, serious damage or serious disruption to the life of the community.

Building on the 1986 statute, the Crime and Disorder Act 1998 (introduced by the Labour Government of Tony Blair) created the Anti-Social Behaviour Order (ASBO) that can be given to an individual who has acted 'in a manner that caused or is likely to cause harassment, alarm or distress'. An ASBO stipulates certain conditions that must be met, and infringing them leads to a criminal record (Pickard 2014b). Subsequently, the Antisocial Behaviour Act 2003 (part 4, section 30) allows for the 'dispersal of groups' if 'any members of the public have been intimidated, harassed, alarmed or distressed as a result of the presence or behaviour of groups of two or more persons in public places in any locality in his police area'. Thus, protests can be curtailed by the police on the basis that they constitute antisocial behaviour. Moreover, under Section 50 of the Police Reform Act 2002, an offence is committed if an individual does not provide his/her name and address when a police officer reasonably believes he/she has engaged in antisocial behaviour. The Conservative–Liberal Democrat coalition government went further with the Antisocial Behaviour, Crime and Policing Act 2014 (part 3, section 34) that permits the pre-emptive dispersal of groups for the purpose of removing or 'reducing the likelihood' of members of the public in the locality being harassed, alarmed or distressed. A person who does not comply is committing an offence liable to 'imprisonment for a period not exceeding 3 months'. This loosely worded legislation sets a lower threshold for the police to disperse groups (Pickard 2014b) and, since being enacted, has also been applied to protest situations as a means of governing them.

Another significant piece of legislation pertaining to the regulation of protests is the Serious Organised Crime and Police Act (SOCPA) 2005 (part 3, sections 110 and 111) that allows police to only need 'reasonable grounds' to arrest someone. Following an arrest, it is for the Crown Prosecution Service to decide whether the case merits going to court. The police use pre-charge bail with exacting conditions attached (introduced by the Police and Criminal Evidence Act 1984) which can forbid an individual from taking part in further protest, for extended periods (sometimes years), despite not having been prosecuted, while the police gather additional evidence in order to be able to press charges, leaving the individual in a 'legal limbo'. Beyond the significant constitutional issue of British law being founded on the principle of 'innocent until proven guilty', pre-charge bail restrictions on someone not actually charged with any crime can include not using the Internet, not associating with certain persons, not going to a particular area, not leaving the country or not participating in further protests. The United Nations Special Rapporteur on the Rights to Freedom of Peaceful Assembly and of Association is highly critical of these measures:

> The use of pre-emptive measures – verbal warnings and arrests – by the authorities against individuals suspected of being likely to commit offences

during protests is troubling. I am also dismayed about very strict police bail conditions which have been imposed on protesters who have been arrested, to deter them from further exercising their rights. Such conditions may be challenged before a court, but the process is costly and can be a strain to some.

(UN 2013: 3)

Following the mass arrests during the student protests of 2010, only a very small proportion of the young people who were arrested and subsequently put on bail ended up actually being sentenced (Rawlinson 2014). This suggests indiscriminate arrest, repressive arrest to hinder further participation in protests and deterrent arrest (and sentencing) to discourage others. This came when legal aid (public funding of legal representation for people on a low income) was being cut as part of government austerity measures, making it financially challenging or impossible for some young defendants to have legal representation.

SOCPA 2005 (sections 132–138) also restricted protest around Westminster. These sections were repealed by the Police Reform and Social Responsibility Act 2011, which created tough new limitations on peaceful protests in Parliament Square, i.e. in view of the Houses of Parliament, the crucible of British democracy.

Lastly, antiterrorism laws were drawn up, voted and applied to peaceful protesters. The Terrorism Act 2000, followed by the Anti-Terrorism, Crime and Security Act 2001, gave police wider stop-and-search powers. Civil rights activists complained that this legislation was used illegally by the police to stop and search peaceful protesters as part of surveillance tactics (see below). In 2010, the coalition government also strengthened antiterrorism measures to curtail perceived threats of 'domestic extremism', including those associated with protest (Jones 2011).

The anti-capitalist and pro-social justice movement Occupy London (an offshoot of Occupy Wall Street) set up camp outside St Paul's Cathedral near the Stock Exchange on 15 October 2011 with the support of the tax-avoidance protest group UK Uncut. They were protesting notably about the links between the state and the world of finance against the backdrop of the recent bailout of banks and financial institutions. The City of London Corporation applied for an injunction against the continuation of the occupation that was granted by Justice Lindblom on 18 January 2012. An appeal was refused on 22 February 2012, which led the following week to bailiffs evicting during the night the remaining protesters (most of whom left peacefully) and removing the tents and other equipment. They were aided by the City of London police and by officers from the Metropolitan Police Service (MPS) (some in riot gear) who made 20 arrests of protesters on a makeshift barricade to resist eviction. Subsequently, police 'angered Occupy London activists after listing the movement among terrorist groups in an advisory notice sent to the business community in the City' (Malik 2011). Occupy also featured in a City of London police PowerPoint presentation on terrorism awareness for use in primary schools. On the slide entitled 'History in City of Terrorism and Domestic issues', featuring Al-Qaida and the Irish Republican Army (IRA), there was a photograph of Occupy London tents to illustrate 'domestic extremism' (Quinn 2015). In this way, police conflated peaceful protest with terrorism and disseminated this to the general public.

Thus, recent laws enacted to govern protest have created a legislative amalgam of peaceful protest, antisocial behaviour and domestic extremism with laws being used to curtail protests and protesters and to dissuade participation in dissent. Further means of dissuading potential protesters are overt and covert surveillance strategies, as shown in the next section.

Data collection as a means of state surveillance of protesters

In recent years, the British police have been making greater use of surveillance and data collection of protests and protesters, followed by the storing of the intelligence on searchable databases. According to the Home Office (2013b), there are four main nationwide databases as part of the Home Office Technology Strategy: the Police National Computer (PNC), Ident1, the National DNA Database (NDNAD) and the Police National Database (PND). These are:

> making available to law enforcement agencies hard (i.e. facts) and soft (i.e. intelligence) information about suspected and actual criminals so the right information is in the right place at the right time. Data is biographic and biometric (facial, finger and DNA).
>
> (Home Office 2013b)

The London Metropolitan Police also has the Crimint (criminal intelligence) database that was created in 1994.

The MPS Commissioner, Paul Stephenson, ordered a review of the policing practices employed in the 2009 youth-led protests that led to several reports, including Her Majesty's Inspectorate of Constabulary's Adapting to Protest (HMIC 2009a), Nurturing the British Model of Policing (HMIC 2009b) and Policing Public Order (HMIC 2011). The High Court of Justice (HCJ) subsequently ruled on 14 April 2011 that the Metropolitan Police should not have employed the containment strategy (as other options were available), that in doing so the police had used 'unnecessary force' and that officers were not 'trained for crowd control operations' (HCJ 2011: §64). The court ruled 'the police may only take such preventive action [containment], as a last resort catering for situations about to descend into violence' (HCJ 2011: §56). These events brought greater public attention to certain policing methods that were employed during protests and especially their practice of 'containment'. Containment involves the encirclement of people with a police cordon to prevent them from leaving an area, which is called informally 'kettling' (Pickard 2014b). Officially, the tactic is used to seal off a small or large group of people who the police consider are causing or are about to cause a serious breach of the peace.

However, during the student-led protests of winter 2010, the police used containment on several occasions in different cities, when they surrounded groups of people (some were protesters, some were bystanders and some were minors) and prevented them from leaving (Sugar 2011; Pickard 2014a). Most notably, police contained the public for up to nine hours in Trafalgar Square and Parliament Square, followed by Westminster Bridge on 9 December 2010 (Penny 2010). Containment was hence further condemned by various official bodies, including the Joint Committee on Human Rights of the British Parliament that denounced its indiscriminate nature and the large numbers of people affected (JCHR 2011). Three school pupils challenged the use of containment of children in the 2010 student-led protest. They sought a Judicial Review in the High Court, arguing it broke the laws of the European Convention on Human Rights (Council of Europe 1950), the United Nations Convention on the Rights of the Child (UN 1989) and the Children Act 2004, regarding the right to protest and the safety of children. The United Nations Special Rapporteur criticises containment in Britain on constitutional grounds: 'I believe that this practice is detrimental to the exercise of the right to freedom of peaceful assembly due to its indiscriminate and disproportionate nature' (UN 2013: 3).

Beyond the police carrying out containment as a means of controlling and curtailing protest *in situ* and *in futuro*, kettling is also used as a method of acquiring information on protesters, which enables the collecting of data and the building of personal profiles of protesters on databases and surveillance. Indeed, the United Nations Special Rapporteur reported that 'it appears that kettling is used for intelligence gathering purposes, by compelling those kettled to disclose their name and address as they leave the kettle, increasing the chilling effect it has on potential protesters' (UN 2013: 3).

Moreover, if someone does not comply and does not provide personal details, he/she could be arrested and given strict bail conditions. Netpol (Network for Police Monitoring) – an activist and campaigning network that monitors the policing of public order and protests – is also highly critical of this practice, considering it a constitutional risk: 'Any power which allows the police to "round up" people engaged in political protest in order to demand their names and addresses under threat of arrest, is a serious and fundamental threat to human rights and freedoms' (Netpol 2013). Indeed, it was ruled an illegal practice by the High Court on 18 June 2013 in the test case of Mengesha vs Commissioner of Police for the Metropolis (HCJ 2013). Therefore, while the Police and Criminal Evidence Act 1984 (section 64A) allows the police to photograph individuals on the street in certain circumstances, the police cannot oblige an individual to identify himself/herself (give his/her name and address) and/or have his/her photograph taken, or be filmed as a condition of leaving a contained area (a kettle) (Liberty and NUS 2014: 13). However, young protesters may not necessarily be aware of their rights and/or might be frightened by the prospect of being arrested if they do not comply.

The filming, photographing and audio-recording of protesters (as well as non-protesting bystanders) during marches and other protest events is allowed by the law and it is generally carried out by Forward Intelligence Teams (FITs). These are police officers who collect intelligence through overt surveillance. Information collected is entered into the searchable Crimint (criminal intelligence) database that holds information on criminals and suspected criminals, which is used daily by police staff (Lewis and Vallée 2009). Thus, convicted criminals and peaceful protesters who have not been arrested are assimilated. The Court of Appeal ruled in 2009 that the material obtained by the FITs and other police officers cannot be systematically kept by the police and that it can only be stored on a case-by-case basis on the Crimint database.

The fingerprinting of protesters constitutes another means of state surveillance in operation in Britain. A majority of police forces now use fingerprint scanning.[2] Fingerprints can be taken with a handheld fingerprint scanner using MobileID (mobile identification) technology, which crosschecks in minutes via satellite with the tens of millions of fingerprints in the searchable finger and palm print PND: Ident1. Police can take fingerprints with or without consent if they 'reasonably suspect' someone has committed a serious offence. According to the police, the fingerprints are not retained and such methods are a more cost-effective and efficient use of police time. For detractors, it is part of large-scale surveillance and monitoring of the population, especially young people and ethnic minorities.

The government also has a DNA database containing biometric information details that was created in 1995; it is the biggest database in the world in relation to the size of the population, and the amount of information held. Following a test case, the Protection of Freedoms Act 2012 (sections 1–25) legislated that 'only people convicted of an offence will have their fingerprint records and DNA profiles retained indefinitely' (Home Office 2013a). Nonetheless, according to the Home Office, as of December 2016, the NDNAD held biometric samples on an estimated 5.2 million individuals; these are broken down into gender, 'ethnic appearance' and age. Importantly, 33.21 per cent of individuals were aged 16–24 when the information was loaded onto the database (Home Office 2017).[3]

Moreover, the police can stop and search an individual without any grounds for suspecting them due to the Criminal Justice and Public Order Act 1994 (section 60) and the Serious Organised Crime and Police Act 2005 (part 3, section 115), while partaking in peaceful protest. Stop and search is a controversial practice in Britain where citizens do not have identity cards and where Black and Asian young men are subject disproportionately to stop and search. In protest situations in London, stop and search is generally carried out by the Territorial Support Group (TSG), which is a central operations unit of the MPS with officers in paramilitary uniforms (helmets and shields) who specialise in public order containment and counterterrorism. A recent Independent Police Complaints Commission (IPCC) report on grievances against the TSG expressed concern on the low success rate of stop and search, as well as 'questionable' reasons for conducting stop and search (IPCC 2015: 14), while the UN Rapporteur declared it should never be used against peaceful protesters (UN 2013: 3).

Lastly, the coalition government voted legislation to enable higher levels of state surveillance of individuals and groups through the greater use of data collection and data storage. Two laws were passed in quick succession that represent a significant development in the broadening of the repertoire of protest governance. The then Home Secretary Theresa May introduced the Data Retention and Investigatory Powers Act 2014 that required internet and telephone companies to retain their communications data for 12 months and it regulated how the police and intelligence agencies could access the information. A legal challenge ensued from Conservative MP David Davis and Labour MP Tom Watson who argued that the statute permitted the police and security services 'to spy on citizens' without sufficient safeguards. Subsequently, the legislation was ruled unlawful by the High Court that found it 'inconsistent with European Union law', i.e. article 8 of the European Convention on Human Rights that protects the 'right to respect for private and family life', as well as the protection of personal data. Therefore, the government introduced the Counter-Terrorism and Security Act 2015 (part 3, section 21, 3b) that allows, amongst other things, for 'relevant internet data' (internet access service or an internet communications service) to 'be used to identify, or assist in identifying, which internet protocol address, or other identifier, belongs to the sender or recipient of a communication (whether or not a person)'. In this way, the legislation permits covert surveillance of political activists and data collection to be undertaken by the state.

Thus, the police have multiple searchable databases holding personal information on innocent protesters obtained by overt and covert means, while the intelligence is stored without the knowledge of the individuals.

The militarisation of the policing of protesters and protest management

The traditional approach to policing in Britain – policing by consent – stems from when Robert Peel founded the Metropolitan Police in 1829. However, in 2014, Liberty, the independent civil liberties organisation, observed a shift towards 'a more militaristic style of policing', associated with what it describes as the 'inglorious reputation for heavy-handed policing and abusive and unlawful use of their powers' (Liberty 2014: 10). Regarding the policing of protests, there appears to have been a move from policing by consent to policing by discouragement.

Three notable allegations were made about the police and their tactics in April 2009 in relation to the G-20 Summit protests in London. First, that the police disbanded the peaceful Climate Camp by force, using the Public Order Act 1986 (section 14) pertaining to 'imposing conditions on public assemblies'. Second, a police officer hit Ian Tomlinson, an innocent bystander not involved in the protest, which led to his collapse and death; an official inquest concluded the newspaper seller was unlawfully killed by a police constable who was later dismissed for gross misconduct. Third, that the police used aggressive crowd control methods on peaceful protesters on the street, including containment (see above).

The police have been 'upping the ante' in terms of their policing strategies exerting violence against protesters. In recent years, police were able to resort to three methods to govern dissent that were previously only deployed by British soldiers (security forces) in Northern Ireland during the 'Troubles' associated with terrorism in the 1970s, but never used on mainland Britain.

The first tactic was the use of CS gas (tear gas) that was deployed for the first time on the mainland during the 1981 riots in Toxteth, Liverpool. Significantly, a London Metropolitan Police officer used CS gas during a UK Uncut protest on Oxford Street, central London, in January 2011. Following complaints, a 2011 report by the IPCC declared that the police officer sprayed peaceful protesters in the face at close range within a crowded area. The IPCC decided that this was in breach of police rules and there was 'insufficient justification' for the officer's actions, as there was no evidence to suggest that those sprayed posed a threat (Taylor 2013).

The second tactic was the authorised use of rubber bullets and plastic bullets or baton rounds – attenuating energy projectiles (AEPs). Following the urban riots of August 2011, the HMIC reviewed their policing tactics (Travis 2011) and as a result specialist CO19 firearms police officers were authorised to deploy AEP if they came under attack, or if there was 'extreme disorder' (HMIC 2011b). Significantly, this authorisation came days before a student protest planned for 9 November 2011, highlighting the conflation between serious disorder and protest, as well as the effort to dissuade young people from taking part in the marches. In a striking acknowledgement of the potentially fatal consequences of AEP, we can read the following on the College of Policing website (2017): 'The AEP is used to dissuade or prevent a potentially violent person from their intended course of action, thereby neutralising the threat. AEP exists as one of the approved less lethal options available in the UK [...].'

The third tactic was a new policing tool that came into being in 2014, when three 20-year-old second-hand water cannons were bought from the German federal police for the MPS by Boris Johnson, Conservative Mayor of London (from 2008 to 2016). He wrote to Home Secretary Theresa May early that year to say he was 'broadly convinced of the value of having water cannon available to the MPS' (Johnson 2014a). When addressing the Police and Crime Committee on 27 February 2014 on the appropriateness of buying and using water cannon in London, the Mayor of London declared the need for the MPS to 'get medieval' in certain public disorder situations (Johnson 2014b: 73). This purchase was made at the request of the Commissioner of the Metropolitan Police, Sir Bernard Hogan-Howe and the Association of Chief Police Officers (ACPO), who warned they expected water cannon to be required because 'the ongoing and potential future austerity measures are likely to lead to continued protest' (ACPO 2014). The deputy prime minister, Liberal Democrat Nick Clegg, expressed disapproval of the procurement of the water cannon: 'I personally think it also rubs up against the long tradition of policing by consent on London's streets' (BBC 2014). Similarly, Liberty considers that the 'water cannon is a symbol of inflammatory, militaristic and brutal policing' (Liberty 2014: 12).[4]

Clearly, the police use of CS gas, the permission to use AEP and the purchase of the water cannon in the governance of protest reflect tactics generally used in the governance of serious public disorder riot situations.

Other policing methods designed to manage and dissuade protest involve electric weaponry. Tasers were introduced into British policing in 2003, as a non-lethal alternative to firearms for police officers faced with potentially dangerous suspects. Police officers are required to take a training course before being allowed to use a Taser and they are instructed to deploy them only when threatened with violence. Tasers have been used in various protest situations, notably at Warwick University in 2014. The Home Secretary, Amber Rudd, authorised in March 2017 more powerful Tasers ×2 (that administer two shots rather than one) to be issued to police officers in England and Wales.

Lastly, other policing methods employed as a governance strategy to manage protest pertained to the dispersal and forced removal of protesters. This is done, for example, by mounted police, as was the case when police on horseback charged a crowd of protesters who had not dispersed when instructed to do so, on 24 November 2010,[5] or by the police and bailiffs during the forced eviction of the Climate Camp in 2009 (see above), Occupy London in 2012 (see above) and environmentalists in 2011 when police used force to remove anti-fracking protesters, including Caroline Lucas (Green Party of England and Wales, MP) who was arrested on 25 September 2013 for wilful obstruction of a public highway and breaching a public order under section 14 of the Public Order Act 1986. She and the other co-defendants were later cleared of all charges.

The shift in policing tactics to a more military-style strategy generally employed in riot situations, as opposed to peaceful protest, no doubt scared some young people from exercising their democratic right to protest.

Conclusion

Britain has experienced an increase in youth-led protest in the twenty-first century. The issues raised by young protesters in Britain and the tactics they employ mirror to a considerable degree those of other young activists elsewhere in the twenty-first century. They constitute part of a bigger whole of political awareness and dissent amongst young people in Europe and around the world, illustrating how young people are re-generating politics in times of crises (Pickard and Bessant 2017).

In parallel, there has been an increase in the governance of dissent in Britain in recent years, with successive governments being responsible for a shift towards more restrictive legislation, more monitoring of protesters and more militarised policing.

While governments lament increasingly about the lower rates of electoral forms of political participation of young people and governments express the will for young people to be more politically engaged (Gallant *et al.* 2018), they have been curtailing freedom of expression and peaceful protest. In legislation and in policing, the boundaries between peaceful protest, potential antisocial behaviour, criminal activity and domestic terrorism have been blurred and state responses conflated. Together, these have a dissuasive effect on participation in protests.

Numerous complaints have been filed and official challenges have been made in the High Court, resulting in the withdrawal of certain restrictive measures, but they have been replaced by other repressive laws. As the United Kingdom leaves the European Union, it is likely that the Conservative Government will replace the Human Rights Act 1998 with a British Bill of Rights and Responsibilities, which would further undermine, stifle and punish young people wishing to exercise freedom of speech and participate in peaceful protest.

Notes

1 The main youth-led organisations and networks included the following: a) The National Union of Students (NUS) is the confederation of the vast majority of further and higher education student unions in the United Kingdom; b) The National Campaign against Fees and Cuts (NCAFC) is an activist network for students and education workers created in February 2010 to campaign against university tuition fees and cuts to the education budget; c) UK Uncut is a movement created in October 2010 to protest against government cuts to public spending and corporate tax avoidance; d) Occupy London is part of the global Occupy movement in favour of democracy and social justice; it had encampments in the capital from October 2011 to June 2012; e) Anonymous UK is part of the international activist and hacktivist Anonymous network; f) The UK Youth Climate Change coalition (UKYCC) is an organisation for young people campaigning against climate change, formed in June 2008.
2 Police in the UK have used LiveScan since 1998. Project Lantern tested and trailed fingerprint scanning in certain police forces starting in 2006. Mobile Identification at Scene (Midas) was rolled out across the UK from 2010 onwards. The London Metropolitan Police adopted handheld devices in 2012.
3 Ages 10–15 years, 8.41%; 15–17 years, 6.75%; 18–20 years, 13.08%; 21–24 years, 13.38%; 25–34 years, 24.5%; 35–44 years, 18.34%; 45–54 years, 9.95%; 55–64 years, 3.89%; 65+ years, 1.36%.
4 The (Conservative) Home Secretary at the time, Theresa May, blocked the use of water cannon on the British mainland in July 2015 and the next London Mayor, Sadiq Khan (Labour), declared he would sell them and spend the money on youth services (Gayle 2016).
5 Official police denials were proved wrong by an amateur phone film (Gabbatt and Lewis 2010).

References

Association of Chief Police Officers (ACPO). (2014). *Water cannon. Use of water cannon by police in relation to disorder incidents and planned public order operations*. Available from www.ipcc.gov.uk/sites/default/files/Documents/speeches/Water_Cannon_Consultation_Response_to_MOPAC_February2014.pdf) [accessed 16 March 2015].

BBC News. (2014). 'Nick Clegg: Water cannon use "against policing by consent"' [online], 12 June. Available from www.bbc.com/news/uk-england-london-27810729) [accessed 16 March 2015].

Bloom, C. (2012). *Riot city. Protest and rebellion in the capital*. Basingstoke: Palgrave Macmillan.

College of Policing. (2017). *Armed policing. Attenuating energy projectiles*. Available from www.app.college.police.uk/app-content/armed-policing/attenuating-energy-projectiles [accessed 24 April 2017].

Council of Europe. (1950). *Convention for the protection of human rights and fundamental freedoms protocol*. (European Convention on Human Rights).

Cunningham, S. and Lavalette, M. (2004). 'Active citizens' or 'irresponsible truants'? School student strikes against the war. *Critical Social Policy*, (24)2: 255–269.

Gabbatt, A. and Lewis, P. (2010). 'Student protests: video shows mounted police charging London crowd'. *Guardian*, 26 November.

Gallant, N., Pickard, S. and Van de Velde, C. (2018). Les politisations actuelles des jeunes adultes. *Lien Social et Politiques*, 80, Spring.

Gayle, D. (2016). 'Water cannon bought by Boris Johnson to be sold off without being used'. *Guardian*, 1 July.

Grasso, M. (2017). Young people's political participation in Europe in times of crisis. In: S. Pickard and J. Bessant (eds), *Young people re-generating politics in times of crises*. London: Palgrave Macmillan, Chapter 10.

Her Majesty's Inspectorate of Constabulary (HMIC). (2009a). July. *Adapting to protest*. London: HMIC.

Her Majesty's Inspectorate of Constabulary (HMIC). (2009b). November. *Nurturing the British model of policing*. London: HMIC.

Her Majesty's Inspectorate of Constabulary (HMIC). (2011a). February. *Policing public order*. London: HMIC.

Her Majesty's Inspectorate of Constabulary (HMIC). (2011b). November. *The rules of engagement*. London: HMIC.

High Court of Justice (HCJ). (2011). *The Queen on the application of Joshua Moos and Hannah McClure v. The Commissioner of Police of the Metropolis [2011] EWHC 957 (Admin) (14 April 2011)*. England and Wales High Court (Administrative Court) Decisions. Available from www.bailii.org/ew/cases/EWHC/Admin/2011/957.html) [accessed 24 April 2017].

High Court of Justice (HCJ). (2013). *Susannah Mengesha vs Commissioner of Police of the Metropolis*. England and Wales High Court (Administrative Court).

Home Office. (2013a). *Protection of Freedoms Act 2012: How DNA and fingerprint evidence is protected in law*. Available from www.gov.uk/government/publications/protection-of-freedoms-act-2012-dna-and-fingerprint-provisions/protection-of-freedoms-act-2012-how-dna-and-fingerprint-evidence-is-protected-in-law) [accessed 17 April 2017].

Home Office. (2013b). *Home Office technology strategy*. Available from www.gov.uk/government/publications/home-office-technology-strategy/home-office-technology-strategy [accessed 17 April 2017].

Home Office. (2017). *National DNA database statistics*. Available from www.gov.uk/govern
ment/publications/home-office-technology-strategy/home-office-technology-strategy
[accessed 17 April 2017].

Independent Police Complaints Commission (IPCC). (2015). *Metropolitan Police Service Territorial Support Group: A review of complaints data and IPCC cases 2008–2012*. Available
from www.statewatch.org/news/2013/jan/uk-territorial-support-group-report.pdf) [accessed
17 April 2017].

Johnson, B. (2014a). Letter from the Mayor of London to the Home Secretary. 6 January.
Available from www.london.gov.uk/sites/default/files/Letter%20from%20Mayor%20
Johnson%20to%20Home%20Secretary%252C%206%20Jan%202014_0.pdf) [accessed
17 March 2015].

Johnson, B. (2014b). Address to Police and Crime Committee. 27 February. Available from
www.london.gov.uk/moderngov/documents/g4999/Public%20reports%20pack%20
Thursday%2027-Feb-2014%2010.00%20Police%20and%20Crime%20Committee.
pdf?T=10) [accessed 17 March 2015].

Joint Committee on Human Rights (JCHR). (2011). *Facilitating peaceful protest*. London:
House of Commons. Available from https://publications.parliament.uk/pa/jt201011/
jtselect/jtrights/123/12312.htm) [accessed 24 April 2017].

Jones, J. (2011). 'The Met turned me into a domestic extremist – with tweets and trivia'.
Guardian, 16 June.

Lewis, P. and Vallée, M. (2009). 'Revealed: police databank on thousands of protesters'.
Guardian, 6 March.

Liberty and National Union of Students. (2014). *Protest. Your rights*. Available from www.
liberty-human-rights.org.uk/sites/default/files/Liberty%20rights%20document%20
A5%2020pp%20FINAL%20web.pdf [accessed 24 April 2017].

Liberty. (2014). *Liberty's response to the Greater London Authority's public engagement on
police use of water cannon*. London: Liberty.

Malik, S. (2011). 'Occupy London's anger over police "terrorism" document'. *Guardian*,
5 December.

Netpol. (2013). 'Protest treated as anti-social behaviour' [online]. 6 May. Available from
https://netpol.org/2013/05/01/protest-treated-as-anti-social-behaviour [accessed 16
March 2015].

Olcese, C. and Saunders, C. (2014). Students in the Winter Protests: still a new social
movement? In: S. Pickard (ed.), *Higher education in the UK and the US: converging
university models in a global academic world?* Leiden: Brill, 250–273.

Penny, L. (2010). 'It was no cup of tea inside the Whitehall police kettle'. *New States-
man*, 25 November.

Pickard, S. (2014a). Productive protest? The contested higher education reforms in
England under the Coalition Government. In: Avril, Emmanuelle and Neem, Johann
(eds), *Democracy, participation and contestation: Civil society, governance and the future of
liberal democracy*. London: Routledge, 93–106.

Pickard, S. (2014b). Keep them kettled! Protesting, policing and Anti-social behaviour
in the student higher education demonstrations. In: S. Pickard (ed.), *Anti-social
behaviour in Britain: Victorian and contemporary perspectives*. Basingstoke: Palgrave
Macmillan, 77–91.

Pickard, S. (2015). The pleasures and problems of political resistance for young people in
contemporary Britain. Journal of Youth Studies Conference, March-April, Copen-
hagen, Denmark.

Pickard, S. (2017). Politically engaged leisure: The political participation of young people in contemporary Britain beyond the Serious Leisure Model. 'The cultures and politics of leisure'. *Angles* 5 November. Available from http://angles.saesfrance.org.

Pickard, S. (2018a). [Forthcoming.] *Politics, protest and young people. Political participation and dissent in Britain in the 21st century.* Basingstoke: Palgrave Macmillan.

Pickard, S. (2018b). [Forthcoming.] The nature of environmental activism among young people in Britain in the twenty-first century. In: B. Prendiville and D. Haigron (eds), Political ecology and environmentalism in Britain. Cambridge Scholars Publishing.

Pickard, S. and Bessant, J. (eds). (2017). *Young people re-generating politics in times of crises.* Basingstoke: Palgrave Macmillan.

Quinn, B. (2015). 'City of London police put Occupy London on counter-terrorism presentation with al-Qaida'. *Guardian*, 19 July.

Rawlinson, K. (2014). 'Revealed: police using pre-charge bail to muzzle protesters'. *Guardian*, 25 December.

Sugar, P. (2011). On riots and kettles, protests and violence. In: D. Hancox (ed.), *Fight back! A reader on the winter of protest.* London: openDemocracy, 77–80.

Taylor, M. (2013). 'Police watchdog criticises Met officer over use of CS gas on protesters'. *The Guardian*, 8 August.

Travis, A. (2011). 'Riot report says police should consider using plastic bullets and water cannon'. *Guardian*, 5 December.

United Nations (UN). (1989). *Convention on the rights of the child.* Adopted and opened for signature, ratification and accession by General Assembly resolution 44/25 of 20 November 1989 entry into force 2 September 1990, in accordance with article 49. Available from www.ohchr.org/en/professionalinterest/pages/crc.aspx [accessed 16 September 2010].

United Nations (UN). (2013). *Statement by the United Nations Special Rapporteur on the rights to freedom of peaceful assembly and of association at the conclusion of his visit to the United Kingdom.* Office of the Higher Commissioner for Human Rights. 23 January. Available from www.ohchr.org/en/NewsEvents/Pages/DisplayNews.aspx?NewsID=12945&LangID=E [accessed 16 March 2015].

Legislation

Police and Criminal Evidence Act, 1984
Public Order Act, 1986
Criminal Justice and Public Order Act, 1994
Human Rights Act, 1998
Crime and Disorder Act, 1998
Terrorism Act, 2000
Anti-Terrorism, Crime and Security Act, 2001
Police Reform Act, 2002
Anti-Social Behaviour Act, 2003
Children Act, 2004
Serious Organised Crime and Police Act, 2005
Police Reform and Social Responsibility Act, 2011
Protection of Freedoms Act, 2012
Anti-Social Behaviour, Crime and Policing Act, 2014
Data Retention and Investigatory Powers Act, 2014
Counter-Terrorism and Security Act, 2015

Part III

Antiterror legislation and the youthful other

Chapter 7

Surveillance of young Muslims and counterterrorism in Kenya

Fathima Azmiya Badurdeen

> On 3 February, the Muslim youth rioted for the second day against a raid on a mosque (Masjid Musa) in Majengo, Mombasa. The youth hurled stones at the police shouting 'release our brothers', which referred to more than 100 people arrested after the mosque raid on Sunday the 2nd of February. The Kenyan police fired tear gas and rubber bullets to disperse the rioting youth.
>
> (Akwiri 2014)

Introduction

The closure of the 'Masjid Musa' mosque resulted in demonstrations and the mobilisation of protest movements in the coastal city of Mombasa, Kenya. The police raid and closure of the mosque were amongst the many counterterrorism strategies intended to curb the al-Shabaab radicalisation[1] occurring on mosque premises in Kenya's coastal region. The Masjid Musa mosque was known to be led by the radical preacher Sheikh Aboud Rogo. Reportedly, the mosque was used for training and recruitment for al-Shabaab (*The Star* 2014a). This government counterterrorism response had come after the Westgate Mall terror attack in Nairobi where 67 people were killed (Howden 2013). The Muslim community viewed this strategy as a heavy-handed tactic that targeted young Muslims, which would further propel anger and violence (Mwakio and Jamah 2014).

Against this backdrop of young Muslims opposing the government's heavy-handed counterterrorism tactics, this chapter examines the impact of state surveillance on 'Muslim youth as suspect communities' in Kenya, as well as the ways these communities mobilise against such surveillance practices and the likely effect of this on open debate and participation in deliberative democratic practice. This is achieved by asking two questions: How have young Muslims from 'suspect communities' mobilised against what they see as the injustices of state surveillance in the Kenyan war on terror? How has government surveillance of 'suspect communities' and their mobilisation efforts affected the power relationships between the targeted communities and the government?

To answer these questions, this chapter is divided into four sections. The first section focuses on political dissent by the suspect communities directed at state surveillance. The second sets out to explain the mobilisation of 'suspect communities' in response to the Kenyan state's 'war on terror', as well as document the state's justification of surveillance. The third section explores how surveillance affects power relationships between a community identified as suspect and the government. The final section considers how state surveillance of 'suspect communities' can be counterproductive – namely, how it works to keep young people from engaging in democratic practices to facilitate relationship-building and from strengthening community responses to terrorism.

This chapter is based on research field notes and 182 interviews with key informants including religious leaders, community mobilisers, youth leaders, women leaders, young people labelled by the state as radicalised/terror suspects and the family members of 'radicals/terror suspects'. This research also draws on my own observations of 21 terror suspect court trials, the analysis of audiovisual material from demonstrations and informal discussions with community members.

Surveillance of suspect communities

'The global war on terror' led to a heightened fear of terrorism that sponsored 'the need' to surveil 'suspect communities' across the world. The notion of 'suspect communities' with regard to counterterrorism was well elaborated in Paddy Hillyard's (1993) work on the growth of the secret state and the effects of the Prevention of Terrorism Acts on the Irish community. In that context, the concept was described as the process of identifying a threat, exemplified and legitimised per the dictates of the state (Hickman *et al.* 2012). A suspect community is therefore defined within a policy framework in which a particular community is viewed and treated very differently from the rest of the population with regard to policy, law and practice, since that community is construed as dangerous or as posing a potential risk. Such a policy framework can have detrimental effects and become itself a self-fulfilling prophecy, resulting in intrusive, alienating, heavy-handed policies and policing techniques that stigmatise an entire community (Vermeulen 2014).

Pantazis and Pemberton (2009: 648) defined suspect communities as 'a subgroup of the population that is singled out for state attention as being problematic'. Hence, in its attempt to prevent political violence such as terrorism, the state may label an individual as suspect (despite them not having committed the act) but based on his or her affiliation with a particular suspect community. This can have long-term implications for the psychological well-being of the individual as well as his or her family and friends. Such labelling by the state of suspect individuals and their actions has the potential to be reproduced by a range of actors, including the media, the general public and members of the community under suspicion (Hickman *et al.* 2012). While these counterterrorism laws are ostensibly designed to protect society, they can also facilitate the social construction of communities in ways that further alienate them. Thus, the notion that certain communities or groups of young people are dangerous and outside the law creates new problems for them by stigmatising and demeaning them, which in turn can lead some of them to carry out criminal acts (Ragazzi 2016).

Aside from facilitating hostility, divisiveness, fear, mistrust and prejudice by reinforcing stereotypes and stigmatising particular Muslim communities through ethnic and racial profiling, counterterrorism policies can also shrink the space for political dissent (Hellyer 2007). The forms of dissent can include campaigning for human rights in response to extrajudicial killings carried out in the war on terror. They might also include young Muslims protesting over their right to obtain certain documents, such as identity cards, and to seek job opportunities in marginalised suspected communities in Kenya's peripheral coastal regions. Dissent can also include Muslim youth supporting a political candidate or party regarded as opposed to the state and hence labelled extremist. Hence, any support for the Mombasa Republican Council, a secession movement in Kenya's coastal region, was viewed with suspicion since most supporters hailed from Muslim-dominated regions (Kisiangani and Lewela 2012). Understanding this can revitalise our thinking about whether counterterrorism policies do indeed prevent terror or contain the covert objective of policing or criminalising dissent.

Like many countries that have introduced counterterrorism legislation and programmes, Kenya is facing the unintended consequences of heavy-handed policing practices that stigmatise specific groups, mainly young Muslims (International Alert 2016). While groups of young Muslims have been framed as terror suspects as a result of ethnic profiling (e.g. Somali Muslims or Coastal Muslims), this context has also prevented young Muslims from expressing dissent on issues that affect them and their communities.

Political dissent and suspect communities

In a context where the Kenyan government, as part of its counterterrorism strategy, targeted young Muslims for surveillance, we saw many young people react with mobilising and protests, some of which became riotous. Discriminatory surveillance practices in this respect included profiling young people based on their ethnic or religious origin. It also entailed putting under surveillance places of worship, schools and other areas in which they congregated, imposing a stop-and-search policy and tighter restrictions for those applying for legal documents (such as identity cards and passports) as well as more measures like monitoring digital communications (Amoore and De Goede 2005). These surveillance practices were backed by counterterrorism legislation – namely, the Prevention of Terrorism Act 2012 (Article 19, 2014). They often targeted specific youth communities deemed vulnerable to Islamist extremism or communities suspected of harbouring terrorists (*The Star* 2014b). Young Muslims – such as those from the ethnic Somali community, Coastal Muslims, North Eastern Muslims and Muslims from Eastleigh in Nairobi – have been identified as vulnerable 'youth at risk' in the Kenyan war on terror. Such identification was based mainly on ethnicity (Somali), religion (viewed as followers of 'distorted' Islamic teachings promoting violence in the al-Shabaab ideology) and location (closeness to the border regions – often neglected or marginalised by the central government – in close proximity to Somalia) (Aronson 2013).

In protest against government surveillance measures that were targeting them, many young Muslims began to mobilise and defend their 'communities' against what they saw as moves that further marginalised and stigmatised them (*The Star* 2014b). Descriptions of such actions as 'malpractice by the state' became key themes in these protests. This was highlighted by one young interviewee who explained:

> We have seen young people coming to the streets whenever there were state malpractices [mosque raids and the killing of Muslim clergy] seen as against the [Muslim] community. Youth don't keep quiet. They have the energy to fight back against any discrimination that they feel targets them and their communities.
>
> (Interview with Hassan,[2] religious leader, 26, 2016, Mombasa)

While there were many acts of political dissent on the part of young people, three kinds of action stand out. One was action in protest against the passing of the Prevention of Terrorism Act 2012 (Laws of Kenya 2012), the successor to the Anti-Terrorism Bill 2003. The opposition movement highlighted the widespread rejection of the legislation by a majority of young Muslims. They felt the legislation had an anti-Muslim stance, inspired by the US-led global war on terror immediately after the 11 September 2001 terror attacks in the United States. Demonstrations were staged in Mombasa and Nairobi against the passage of the legislation. Notwithstanding these protests the bill was eventually enacted as 'a solution' to the increasing acts of terror in the country (Nyaundi 2014). In 2014, the Security Amendment Act further strengthened the Prevention of Terrorism Act (Wordpress 2014). These laws served to identify young Muslims as responsible for terrorism, and implied that the closer governance of 'problem youth' was a solution to the problem of terrorism.

Second, young people carried out demonstrations and protest marches in response to the closing of mosques following increased surveillance efforts after a string of terrorist attacks in Kenya. On 2 February 2014, the Masjid Musa mosque in Mombasa had been closed down on the premise that the mosque was in the hands of radicalised or extremist young people who were convening meetings on behalf of al-Shabaab. The mosque closure followed the terrorist attack at the Westgate Shopping Mall, which sparked increased surveillance of public spaces like mosques (*The Star* 2014c). According to activists and media reports, the police used force to stop a meeting held on 2 February 2014, raided the place and arrested 129 people (Sperber 2015).

These protests against the closure of the mosque were ignored. The entire incident was framed as support for radicalism rather than as support for the right of the community to have a place of worship (Field Notes, interfaith summit, 22 June 2016). Following reports by intelligence agencies of more protests after the Friday prayers, there was an increase in police numbers, and a rapid deployment unit was created at the provincial police headquarters in Mombasa (*The Star* 2014c). The closure of three more mosques (Swafaa, Minaa and Sakina) in subsequent raids in November 2014 provoked more resentment amongst many young Muslims (*Daily Nation* 2014). The mosques were closed because of their alleged links with the al-Shabaab network. The police found weapons and al-Shabaab promotional materials on the mosque premises. Along with the mosque closures, hundreds of people were arrested. Young people subsequently launched a series of demonstrations which involved pelting the police with stones and led to riots (Akwiri 2014; Igunza 2014). As one young participant recalled:

> [E]ven I seeing it on TV felt myself being radicalised. It's the anger over how the complete operation [mosque raid] was conducted. The mosque was forcefully closed. The time was the congregation time [time for midday prayers], then you saw young people getting arrested; whatever the reason may be, the way it was conducted was wrong. It only drew more sympathy for these young Muslims as we saw more anger and more resent[ment].
> (Interview with Abubakr, 28, religious leader, 3 March, Mombasa)

Third, was a series of protests by young people who took to the streets to demonstrate against the killing of religious leaders by the police. They viewed these as extrajudicial killings associated with official attempts to counter radicalisation and extremism (International Center for Transitional Justice 2013; Kajee 2014). Notably, amongst these protests were the demonstrations, which took place immediately after the assassination of Aboud Rogo and Ibrahim Rogo Omar. On 31 August 2012, young Muslims protested after the Friday mosque prayers in response to the killing of the religious leader Aboud Rogo, a controversial radical cleric, on 27 August 2012. The young protesters clashed with the police. They wanted answers about who was responsible for the killing. The protests turned violent after some marchers began stoning cars and lighting bonfires (Craig 2012). Further protests took place on 5 October 2013, when young people protested again after Friday prayers against the killing of another cleric, Ibrahim Rogo Omar (BBC News Africa 2013). The police, however, denied any involvement in the deaths of the two clerics.

The two religious leaders were alleged to have been radical preachers, recruiting young people and encouraging them to join al-Shabaab. Muslim clerics and human rights groups condemned the killing of religious leaders and the absence of any accountability (Gaffey 2016). Police also retaliated against young people during the demonstrations against the killings. Unsurprisingly, some young people believed 'the police and the state [were] targeting the Muslim religious leaders [because they had been] defined and constructed by the state narrative as radical preachers' (Interview with Hussein, 27, youth leader, 22 February 2016).

This political dissent by young people reveals their resentment towards the government and its counterterrorism strategies, especially the way they targeted young people from particular Muslim communities.

Contextualising surveillance and suspect communities in the Kenyan war on terror

Kenya has been subject to terrorist attacks, due in part to its strategic geographic positioning in the East African region (Aronson 2013). Since the 1980s, the country has been grappling with terror attacks associated with Islamist extremism, such as when the Palestine Liberation Organization attacked the Jewish-owned Norfolk Hotel in Nairobi. This was followed by other attacks such as the bombings at the United States Embassy in 1998 in Nairobi, the Kikambala Hotel bombing in Mombasa in 2002, the Westgate attack in 2013 (Miller 2013), the Mpeketoni attacks in 2014, the terror attacks in Mandera and the Garissa University attack in 2015 (Pate *et al.* 2015). With the increase in terrorist attacks and the emergence of new 'terrorism tactics', counter strategies were developed involving the close surveillance of young people (Mukinda 2014) as a pre-emptive strategy (Aradau and Van Munster 2007).

The official justification for the surveillance of 'suspect communities' was said to be needed to 'fight terrorism'. Central to the counterterrorism strategies was the surveillance of places of worship and a focus on geographic areas, sometimes referred to as 'hotspots of al-Shabaab radicalisation'. These 'hotspots' were identified by factors such as the number of radical incidents in a specific area, the number of people identified as radicals living in a particular locale, the presence of a radical leader or a mosque suspected of being used for radical preaching, the presence of a madrasa (an educator) suspected of teaching radical views or simply a place with a significant Muslim population (Interview with law enforcement official, Mombasa, 5 March 2016).

While 'the radicals' were identified as individuals within these predominantly Muslim communities, entire communities easily became 'suspect communities'. This was a consequence of those communities being considered to be 'at risk' of radicalisation, of being vulnerable to violent extremism or because they may aid perpetrators (Field Notes, discussion after the Countering Violent Extremism (CVE) Meeting, 21 June 2016). By focusing on the community at large, the government ignored strategies arguably better focused on specific individuals.

These surveillance strategies relied on profiling Muslim individuals and communities from Somali, Coastal or North Eastern communities to produce data on 'suspect communities' (Burbidge, 2015; McConnell 2011; Nzes 2012). Prior to 2011, most attacks in Kenya were affiliated with al-Qaeda, and suspects were mainly of 'Arab origin'. The primary surveillance measures used at that time included the racial profiling of ethnic Arab populations in Nairobi and Mombasa (Rabasa 2009) with routine arrests at airports, detention and stop-and-search measures. This racial profiling drew on similar techniques used in the West to deal with 'terror threats' by focusing on particular suspect communities (Prestholdt 2011; Vittori *et al.* 2009).

Since 2012, the perpetrators of terror attacks have been affiliated with the al-Shabaab Somali militant group. With the entry of the Kenyan Defence Forces (KDF) into Somalia, Kenya saw a host of terror attacks in retaliation by al-Shabaab. In October 2011, the KDF were deployed in Somalia's Juba Valley against al-Shabaab. Known as Operation Linda Nchi [Protect the Country] this was a co-ordinated military operation between the Kenyan and Somali militaries. The operation was approved after a string of cross-border kidnappings of foreigners and aid workers from Kenya. In retaliation for this operation, al-Shabaab carried out terrorist attacks, like the attack at the Westgate Mall in Nairobi and the Garissa University attack (Nzes 2012). As instances of domestic terrorism increased, so too did the government's surveillance measures directed at young Muslims in areas such as Kenya's coastal and eastern regions (Anzalone 2012).

State narratives on the discourse of youth dissent

The government justified its repression of these protests by appeals to a law-and-order agenda and the need to secure peace in the region. The government argued the youth protests in response to the mosque closures were engineered by 'Muslim youth radicals' recruited by their 'radical leaders' or puppeteered by 'outsiders' (e.g. foreign al-Shabaab 'radical entrepreneurs'). Anyone associated with the protest actions was treated as a 'radical element', a portrayal that strengthened the narratives of law enforcement officials that 'these youth are violent and are followers of these preachers'. As 'radicals, extremists, or sympathisers of al-Shabaab', the state was justified in curbing their activities. Indeed, given that framing – in which 'demonstrators were often construed as the bad guys – it just seemed the logical thing to do to use force against these young people because they threatened peace in the region' (Interview with Sauda, 32, community mobiliser, Mombasa, 15 March 2016).

Contemporary counterterrorism discourse and the measures it justified distinguished between 'moderates' and 'radicals'. This act of being labelled by the state served to prevent some young people from participating in any protest activities. In this way, it had a powerful repressive effect because if they exercised their democratic right (as Muslims) to participate in demonstrations then it was said to be equivalent to supporting radicals, being a sympathiser or being in some way aligned with 'al-Shabaab radicalisation' (Kajee 2014). This was highlighted by one young person:

[I]f you get caught in a protest or demonstration, the authorities are not bothered about what your real intentions are. If the issue is demonstrating against the closure of a mosque or mosque raids, our intention is to show that the closure of a mosque, which is our place of worship, is wrong, and the way [means] it is done is wrong. … The authorities [police] view this as a protest against their efforts linked to national security … and our demonstrations are viewed as if we were against national security, rather than … asking for rights pertaining to a place of worship.

(Interview with Ali, 23, Mombasa, 12 February 2016)

The same point was made by a participant in a focus-group discussion: 'Muslims who do not support the closure of mosques in line with the government counterterrorism strategy are automatically viewed as radicals supporting al-Shabaab activities like the Westgate terror attack' (focused group discussion (FGD) with youth in Mombasa, 24 November 2016). Young people who demonstrated against the mosque closures were arrested as their protests were certainly not treated as an expression of basic rights. Osman expressed his feelings about how young Muslims were viewed during the demonstrations: '[W]e [young people] demonstrated [against] the repression of this right, [and] we were seen as radicals and extremists fighting against the state and their countering extremism strategies' (Interview with Osman, 29, Mombasa, 21 April 2016). Mahmoud, a young tuk-tuk driver (a three-wheeled form of public transport), agreed:

[T]he youths arrested during the protests were not supporters of al-Shabaab activities … they were only protesting for their religious rights and that of their place of worship. This is what any Muslim would do … they felt their religion was threatened.

(Interview with Mahmoud, 27, Mombasa, 20 March 2016)

One clear consequence of this labelling of young people was self-imposed censorship or self-governance (Cherney and Murphy 2016). In other words, young Muslims said they feared being labelled as 'radicals'. It mattered to them that they got social acceptance as 'moderates' so as 'to be in the good books' of the authorities and to prevent police surveillance of themselves and their families (Interview with Juma, 23, Lamu, 21 June 2016). Being 'good young Muslims' was equated with being 'moderate Muslims', and conforming to this norm meant supporting government counterterrorism efforts, not being involved in any form of political dissent. This idea of being a 'good guy' was reinforced by political and media rhetoric. As one young person explained:

[S]ome of us don't want to be involved with any discussion or workshop that discusses radicalization or extremism. … [W]e fear how we would be perceived. … [S]ometimes these activities are just to test our affinity. … [O]nce we come onto the police radar, they will follow us and our families.

(Interview with Omar, 24, Mombasa, 21 January 2016)

Surveillance of suspect communities and its implications

The surveillance of 'suspect communities' involves the exercise of state political power over both public and private spaces. Mass surveillance violates the basic human rights of any democratic community to exercise control over its collective practices, as well as breaching individual privacy rights. Knowing that they were under surveillance in public and private spaces was designed to regulate behaviour and conformity to the idea of being a moderate Muslim, as defined by the Kenyan state and the media. It did result in Muslims abstaining from deliberative democratic practices because of the fear of being watched and labelled. For example, participating in media forums led some young people to desist from online commentary for fear of being labelled as radicals. As Hamisi, a university student, explained:

> How many Muslim youth leaders talk on these issues [deconstruction of radical ideologies]? Just a handful. It's because they fear being labelled. As soon as you give your opinion, you are either a radical or a moderate that is a government supporter. … Having the goodwill of the government is important. … We have seen various harassments of youths because of their beliefs…. It's fear; the outcome is not addressing the core issues beneath the radical cover but addressing the naming of the wound – the radical.
> (Interview with Hamisi, 28, youth leader, Kwale, 20 February 2016)

Labelling individuals or issues as 'radical' has also prevented many young people from discussing concerns pertaining to their regions, such as their marginalisation or discrimination against their identity rights. As one participant put it:

> Now tell me … what did the youths do after the raids? They went to the streets to demonstrate their anger. All they wanted was answers … answers to why their mosque was closed. On which radical grounds was it closed? Why were some of their friends arrested? Why were they targeted? Why couldn't they respect the mosque by removing their shoes when they [police] entered? … All they needed were answers. Police said the mosque premises had weapons and radicalised youth. How do the youths around this place believe that, even if it may be right? For youths around this place have been constantly harassed by police. … If the authorities cannot provide tangible answers, it's natural they will demonstrate. This can further fuel their already existing anger.
> (Interview with Khalid, 38, community mobiliser,
> Mombasa, 22 February 2016)

If these questions go unanswered, some young people may turn to alternative political activities that they think will help them tackle the injustices caused by targeting entire Muslim communities (Songok 2014). Incidents like these have the potential to facilitate the radicalisation of Muslim youth. How these issues are dealt with as part of a counterterrorism strategy needs to acknowledge the pivotal role played by the need to build trust in the community (Juma 2015). This was highlighted by one religious leader:

> Our actions give room to those radicals to make use of our youths. They exploit these conditions of the youths' grievances for their cause. Usually they [radicalisers] explain government discrimination of Muslims just like in many other countries in the West. … Youths can be trapped into these narratives.
>
> (Interview with Taheer, 39, religious leader, Mombasa, 11 November 2016)

This highlights the need to ensure that counterterrorism strategies or interventions by law enforcement officials offer convincing justifications to the community for those interventions. Doing this faces some inherent problems. First, counterterrorism interventions that involve military or police actions usually need to be carried out in a timely manner and, typically, they need to be carried out in secrecy. This poses an inherent difficulty in explaining the process and the likely impact. As one participant indicated:

> there is very little information to be revealed before an intervention, including how we work. Most of the interventions are based on intelligence, and we neither have the time nor the possibility to give prior warnings, as it can affect the intervention. … After all, we have to nab the radicalised elements before it's too late, which could threaten the peace of the country.
>
> (Fredrick, law enforcement official, 2 September 2016)

While some amount of secrecy may be needed in effective counterterrorism strategies, convincing arguments and justifications still need to be given to communities. Being unaware of counterterrorism interventions can have disastrous effects on vulnerable communities as they feel the issues are not dealt with properly, and actions are applied in a discriminatory fashion. For example, many thought that law enforcement officials could at least partly justify their behaviours and the strategies they employed to boost co-operation from the community. This could improve the way counterterrorism strategies were viewed and how the after-effects are experienced by the communities involved. The mainstream media have the potential to play a prominent role in disseminating information that can build community support to assuage tensions between law enforcement officials and the community. Furthermore, there is a need for collaboration between the police and the community in designing and implementing counterterrorism policies and practices, such as the government-led homeland security programme, Nyumba Kumi, a community-level policing

initiative. This can help build partnerships where Muslim young p
voluntarily provide lateral surveillance within their respective com
strengthen counterterrorism strategies.

Lateral surveillance in counterterrorism moves from the top-down monitoring of the state on its citizens to a community member–member, or peer–peer monitoring strategy in eliciting information – an approach which internalises and deploys government counterterrorism strategies in the private sphere between people within communities (Andrejevic 2005). Used increasingly in counterterrorism efforts, this strategy has worked well in communities in generating data on suspicious individuals and their activities. However, it comes with its own problems of creating fear and mistrust of 'the other', generating a culture of suspicion (Chan 2008). The Nyumba Kumi initiative in Kenya relies on lateral surveillance within communities. Amidst few successes by this initiative in generating data on suspicious individuals and foiling terror activity attempts, the initiative evokes mixed reactions in terms of effectiveness, as some view it suspiciously as a programme created by the government for spying (Field Notes, CVE Meeting with youth, Mombasa, 21 September 2016).

Finally, there is a need to open up spaces for young people to freely articulate issues concerning themselves and their communities. Youth-led discussions could also facilitate the process of local democratic decision-making (Badurdeen 2014) as most decisions pertaining to young people and their lives are made by elders, with young people serving as mere bystanders (Field Notes, CVE Meeting with youth leaders from universities, Mombasa, 22 September 2016).

While various activities such as 'youth parliaments, youth clubs and other initiatives are evident, there is a need for more inclusive youth-led initiatives that enable young people to deliberate in decision-making processes that affect them and to articulate their grievances' (Wahid, 25, community mobiliser, 20 September 2016). The absence of discussion forums for young people may encourage some to join organisations that fulfil their needs and desires for political engagement and that allow them to air their grievances (Field Notes, 21 September 2016). Digital platforms can be effective in this regard, as many young people are increasingly involved in daily discussions in digital space. These platforms can also be used to open deliberative spaces for political movements to articulate and communicate regional grievances. Having said this, there are reasons for concern that the stringent nature of digital surveillance methods used to enact state counterterrorism policies could constrain and close the digital spaces now used by young people to deliberate and air their grievances (Badurdeen 2016).

Conclusion

Engaging in political dissent via protest, demonstrations as a response to state counterterrorism strategies are a difficult way for Muslim communities to express their grievances. It is bad enough that state surveillance of public places such as mosques, madrasas and schools prevents young people from engaging in open discussion about matters that directly affect them. Some respondents explained that they avoided open discussion on issues pertaining to religion or discrimination against Muslims for fear of the likely repercussions (FGD, young Muslim, Kwale, 3 June 2016).

Surveillance measures that repress open discussion and debate in the public sphere also extend to the digital space. Social media is widely used by young people in Kenya, and so the pervasive monitoring and surveillance of social media inhibited deliberative practices in that space. Constant surveillance by law enforcement officials on websites, Twitter accounts and WhatsApp applications facilitated information gathering of suspected individuals by the state. However, such action also reduced space for debate and changed the ways young people used digital platforms; for example, they were reportedly 'more careful' on the Internet and do not browse certain sites for fear of being tracked or being considered interested in the promotion of 'radical activities' (Badurdeen 2016). As Rahab explained:

> [T]hese days, using the web needs to be done with utmost care. You never know when you will end up on a site branded as an extremist website. ... [B]y merely browsing, you can end up in trouble with the cops [police].
>
> (Rahab, 26, community mobiliser, 12 February 2016)

Online activism is a newly emerging public sphere and efforts to open up this medium for young people is important for those interested in addressing the grievances of young Muslims and engaging them in civic activities (Field Notes, discussion on Social Media Platforms for Youth, 12 December 2016). A better understanding of social media use in this context would help to extend the discussion surrounding the effects of surveillance on young Muslim communities.

Notes

1 Al-Shabaab radicalisation refers to individuals who sympathise with or accept the al-Shabaab ideology. Al-Shabaab (Harakat al-Shabaab al-Mujahideen, also known as 'The Youth' in Arabic) is a militant organisation originating in Somalia. The organisation is known for its extremism and terrorist tactics, having threatened many countries in East Africa. Their recruitment strategies extend to the East African region, like Kenya, where young people are deemed vulnerable to radicalisation as they come from marginalised regions experiencing socio-economic deprivation.
2 Pseudonyms have been used in this chapter.

References

Akwiri, J. (2014). 'Kenyan port city again hit by riots over mosque raid'. Reuters. 3 February. Available from www.reuters.com/article/us-kenya-unrest-mombasa/kenyan-port-city-again-hit-by-riots-over-mosque-raid-idUSBREA120H020140203 [accessed 10 June 2016].

Amoore, L. and De Goede, M. (2005). Governance, risk and dataveillance in the war on terror. *Crime, Law & Social Change*, 43(1): 149–173.

Andrejevic, M. (2005). The work of watching one another: lateral surveillance, risk, and governance. *Surveillance and Society*, 2(4): 479–497.

Anzalone, C. (2012). Kenya's Muslim Youth Center and Al-Shabaab's East African Recruitment. Combating Terrorism Center. Available from www.ctc.usma.edu/posts/kenyas-muslim-youth-center-and-al-shababs-east-african-recruitment [accessed 2 March 2017].

Aradau, C. and Van Munster, R. (2007). Governing terrorism through risk: taking precautions, (un)knowing the future. *European Journal of International Relations*, 13(1): 1–42.

Aronson, S. (2013). Kenya and the global war on terror: neglecting history and geopolitics in approaches to counterterrorism. *African Journal of Criminology and Justice Studies*, 7(1): 24–34.

Badurdeen, F. (2016). 'Digital surveillance in countering cyber terrorism in Kenya: ethical dilemmas and moral panics'. Paper presented at the CODESRIA workshop *The African State and the Public Cyber-Security Service* in Dakar, Senegal on 26 November 2016.

Badurdeen, F. (2014). Youth participation in local governance: youth perspective from the coast region of Kenya. *Africa Peace and Conflict Journal*, 7(2): 89–95.

BBC News Africa. (2013). 'Mombasa riots after Kenyan cleric Ibrahim Omar killed'. 4 October. Available from www.bbc.com/news/world-africa-24398548 [accessed 2 February 2017].

Burbidge, D. (2015). 'The Kenyan State's Fear of Somalia Identity. Conflict Trends-ACCORD'. Available from www.accord.org.za/conflict-trends/the-kenyan-states-fear-of-somali-identity/ [accessed 12 March 2016].

Chan, J. (2008). The new lateral surveillance and a culture of suspicion. In: M. Deflem and J. Ulmer (eds), *Surveillance and governance: crime control and beyond*. West Yorkshire: Emerald Group Publishing Limited, 223–239.

Cherney, A. and Murphy, K. (2016). Being a 'suspect community' in a post 9/11 world. The impact of the war on terror on Muslim communities in Australia. *Australian and New Zealand Journal of Criminology*, 49(4): 480–496.

Craig, J. (2012). 'Kenyans protest killing of radical Muslim cleric'. VOA News, 28 August. Available from www.voanews.com/a/kenyans-protest-killing-of-radical-muslim-cleric-aboud-rogo-mohammed/1496935.html [accessed 2 February 2017].

Daily Nation. (2014). 'Minaa and Swafaa mosques in Mombasa closed after another dawn raid'. *Daily Nation*, 19 November. Available from www.nation.co.ke/counties/mombasa/Two-mosques-in-Mombasa-closed-after-another-dawn-raid/1954178-2527980-jb6otj/index.html [accessed 2 February 2017].

Gaffey, C. (2016). 'Kenyan police have killed or "disappeared" 81 Muslims since 2012'. *Newsweek*, 12 August. Available from www.newsweek.com/kenyan-police-killed-disappeared-81-muslims-2012-report-529620 [accessed 12 February 2017].

Hellyer, H. (2007). *Engagement with the Muslim community and counter-terrorism: British lessons for the West*. Washington DC: The Sabon Center for Middle East Policy at Brookings.

Hickman, M., Thomas, L., Nickels, H. and Silvestri, S. (2012). Social cohesion and the notion of 'suspect communities': A study of the experiences and impact of being 'suspect' for Irish communities and Muslim communities in Britain. *Critical Studies on Terrorism*, 5(1): 89–106.

Hillyard, P. (1993). *Suspect community: people's experiences of the Prevention of Terrorism Acts in Britain*. London: Pluto Press.

Howden, D. (2013). 'Terror in Nairobi: The full story behind al-Shabaab's mall attack'. *Guardian*, 4 October. Available from www.theguardian.com/world/2013/oct/04/westgate-mall-attacks-kenya [accessed 5 March 2016].

International Alert. (2016). *We don't trust anyone: strengthening relationships as the key to reducing violent extremism in Kenya*. London: International Alert.

International Center for Transitional Justice. (2013). *EU decries rise in extra-judicial killings in Kenya*. Available from www.ictj.org/news/eu-decries-rise-extra-judicial-killings-kenya [accessed 10 June 2016].

Igunza, E. (2014). 'Tackling the radical Muslim youth of Mombasa'. BBC News Africa, 19 November. Available from www.bbc.com/news/world-africa-30118119 [accessed 12 March 2017].

Juma, Y. (2015). 'Muslim suspects face rough justice in Kenya's terror war.' *Anadolu Agency*. 10 October. Available from http://aa.com.tr/en/africa/muslim-suspects-face-rough-justice-in-kenyas-terror-war/437466 [accessed 15 September 2016].

Kajee, A. (2014). 'Muslims feel under siege in Kenya'. Al Jazeera. 13 November. Available from www.aljazeera.com/indepth/opinion/2014/11/muslims-feel-under-siege-kenya-201411911737464684.html [accessed 12 June 2016].

Kisiangani, E. and Lewela, M. (2012). 'Kenya's Mombasa Republican Council: liberators or nascent radical fanatics?' Institute for Security Studies. 27 June. Available from https://issafrica.org/iss-today/kenyas-mombasa-republican-council-liberators-or-nascent-radical-fanatics [accessed 21 June 2017].

Laws of Kenya. (2012). *Prevention of Terrorism Act 30 of 2012*. Available from www.kenyalaw.org/lex/actview.xql?actid=No.%2030%20of%202012 [accessed 23 May 2016].

McConnell, T. (2011). 'Kenya: where all Somalis are suspects?' Public Radio International. 4 November. Available from www.pri.org/stories/2011-11-04/kenya-where-all-somalis-are-suspects [accessed 10 June 2016].

Miller, E. (2013). 'Al-Shabaab attack on Westgate Mall in Kenya. Background Report'. Available from www.start.umd.edu/sites/default/files/publications/local_attachments/STARTBackgroundReport_alShabaabKenya_Sept2013.pdf [accessed 12 January 2017].

Mukinda, F. (2014). 'Masterminds target youth to execute terror attacks'. *Daily Nation* [online], 3 March. Available from http://mobile.nation.co.ke/news/Masterminds-target-youth-to-execute-terror-attacks/-/1950946/2228164/-/6w8l65/-/index.html [accessed 5 May 2016].

Mwakio, P. and Jamah, A. (2014). 'Fury over state's mosques closure threat'. *Standard Digital*, 15 September. Available from www.standardmedia.co.ke/article/2000135073/fury-over-state-s-mosques-closure-threat [accessed 12 May 2017].

Nyaundi, K. (2014). *How does the implementation of counter terrorism measures impact on human rights in Kenya and Uganda?* PhD thesis, Department of Public Law, Faculty of Law, University of Cape Town.

Nzes, F. (2012). Terrorists attacks in Kenya reveal domestic radicalization. Combating Terrorism Center. Available from www.ctc.usma.edu/posts/terrorist-attacks-in-kenya-reveal-domestic-radicalization [accessed 10 June 2016].

Pantazis, C. and Pemberton, S. (2009). From the 'old' to the 'new' suspect community: examining the impacts of recent UK counter-terrorist legislations. *The British Journal of Criminology*, 49(5): 646–666.

Pate, A., Jensen, M. and Miller, E. (2015). 'Background Report: Al-Shabaab Attack on Garissa University in Kenya'. START April. Available from www.start.umd.edu/pubs/STARTBackgroundReport_alShabaabGarissaU_April2015.pdf [accessed 10 June 2017].

Prestholdt, J. (2011). Kenya, the United States, and counterterrorism. *Africa Today*, 57(4): 2–27.

Rabasa, A. (2009). 'Radical Islam in East Africa'. RAND Corporation. Available from www.rand.org/content/dam/rand/pubs/monographs/2009/RAND_MG782.pdf [accessed 2 January 2017].

Ragazzi, F. (2016). Suspect community or suspect category? The impact of counter-terrorism as 'policed multiculturalism'. *Journal of Ethnic and Migration Studies*, 42(5): 724–741.

Security Amendment Act, Article 19. (2014). *Kenya: concerns with Security Laws (Amendment) Bill*. 17 December. Available from www.article19.org/resources.php/resource/37800/en/kenya:-concerns-with-security-laws-(amendment)-bill [accessed 2 March 2017].

Songok, J. (2014). 'Garissa SUPKEM demand reopening of mosques in Mombasa'. *Kenya News Agency*. 24 November. Available from http://kenyanewsagency.go.ke/en/garissa-supkem-demand-reopening-of-mosques-in-mombasa/ [accessed 23 May 2016].

Sperber, A. (2015). Little Mogadishu, under siege. *Foreign Policy*. 14 April. Available from http://foreignpolicy.com/2015/04/14/kenya-shabab-somalia-garissa-kenyatta/ [accessed 10 June 2016].

The Star. (2014a). 'Kenya: several youth arrested in Mombasa Swafaa Mosque after forcing their way in for prayers'. *The Star*, 21 November. Available from http://allafrica.com/stories/201411210434.html [accessed 2 January 2017].

The Star. (2014b). 'Muslim leaders caution against Masjid Musa closure, acknowledge the radicalisation problem'. *The Star*, 5 February. Available from www.the-star.co.ke/news/2014/02/05/muslims-leaders-caution-against-masjid-musa-closure-acknowdledge-the_c892016 [accessed 10 June 2016].

The Star. (2014c). 'NIS wants Masjid Musa closed over terrorism'. *The Star*, 7 February. Available from www.the-star.co.ke/news/2014/02/07/nis-wants-masjid-musa-closed-over-terrorism_c893227 [accessed 2 January 2017].

Wordpress. (2014). The Security Laws (Amendment) Act 2014. Available from https://kenopalo.files.wordpress.com/2014/12/the-security-laws.pdf [accessed 2 January 2017].

Vermeulen, F. (2014). Suspect communities – targeting violent extremism at the local level: policies of engagement in Amsterdam, Berlin, and London. *Terrorism and Political Violence*, 26(2): 286–306.

Vittori, J., Bremer, K. and Vittori, P. (2009). Islam in Kenya and Tanzania: ally or threat in the war on terror? *Studies in Conflict and Terrorism*, 32(12): 1075–1099.

Chapter 8

On *becoming* 'radicalised'

Pre-emptive surveillance and intervention to save the young Muslim in the UK

*Vicki Coppock, Surinder Guru
and Tony Stanley*

Introduction

In recent times, young British Muslims have become objects of the official discursive construction of 'vulnerability to radicalisation' (HM Government 2011; 2012). This is shaped by ideologically charged and contentious ideas about how radicalisation allegedly takes place, set alongside equally problematic constructions of childhood, youth and psychological well-being (Coppock and McGovern 2014). Thus, pathologised as 'psychologically vulnerable', young British Muslims are constructed by the state as appropriate objects for pre-emptive surveillance and intervention as 'terrorists in the making'. Moreover, this project of youth governance is bolstered through the mobilisation of dominant administrative discourses of child protection and safeguarding which construct the young British Muslim as simultaneously 'suspect' and in need of being 'saved' (Department for Education (DfE) 2015a, 2015b). In this, public-sector professionals from education, health and child welfare are conscripted into the state's securitisation agenda, effectively working as agents of its internal security services, to surveil and regulate young people's political dissent. In this chapter, we critique and contest the assumptions and arguments contained within this official discourse concerning vulnerability to radicalisation that give legitimacy to statutory surveillance and intervention in the lives of British Muslim children and young people, purportedly *in their best interests*, but which effectively undermine their political agency. Implications for the political rights of young British Muslims will be our central concern, alongside consideration of the implications for practitioners who are expected to deploy these strategies on behalf of the state (DfE 2015b; HM Government 2012, 2015). Our central thesis is that the discourse of vulnerability to radicalisation rests principally upon the dual construction of the young British Muslim as both a politically risky subject and a psychologically vulnerable subject. In this, the risk–vulnerability nexus acts as a powerful mechanism for the governance of Muslim young people's political identities through the normalisation and routinisation of surveillance and monitoring practices, justified as safeguarding their 'childhood' and 'youth' (DfE 2015a, 2015b).

The chapter is divided into three parts. Part 1 provides a contextual overview of the architecture of the UK government's counterterrorism and anti-radicalisation strategy 'PREVENT', which is the foundation for contemporary state securitisation and surveillance practices directed towards the governance of the politics of young British Muslims. In Part 2, we draw on theoretical insights from the sociology of childhood and Ian Hacking's (2006) *Making Up People*, to illustrate and critique the role of 'psy' sciences in constructing the classifications, technologies and practices within PREVENT that facilitate the governance of the political identities of young British Muslims through the identification and correction of potentially 'dangerous minds' (Coppock and McGovern 2014). Finally, in Part 3, we consider the implications of PREVENT for social workers in the UK who are legally required to deploy these technologies and practices in the surveillance and governance of the politics of young British Muslims on behalf of the state.

The politics of securitisation: PREVENT and the governance of political dissent amongst Muslim children and youth

The British government's PREVENT strategy was first introduced in 2003 (revised in 2006 and 2011), with the aim of stopping people 'becoming terrorists or supporting terrorism' (HM Government 2011: 6). It is one of four strands that comprise its overarching 'CONTEST' counterterrorism strategy; the other three being PURSUE (to stop terrorist attacks), *PROTECT* (to strengthen protection against terrorist attack) and PREPARE (to mitigate the impact of a terrorist attack). A key objective of PREVENT is to 'respond to the ideological challenge of terrorism' by undertaking 'counter-ideological work' designed to ensure that there should be 'no ungoverned spaces in which extremism is allowed to flourish' (HM Government 2011: 9). The government defines extremism as 'vocal or active opposition to fundamental British values including, democracy, the rule of law, individual liberty and mutual respect and tolerance of different faiths and beliefs' (HM Government 2011: 107). Radicalisation is understood as 'the process by which a person comes to support terrorism and forms of extremism leading to terrorism' (108). In this, extremist ideas are purportedly propagated and disseminated by extremist activists who 'radicalise' persons who are 'vulnerable' to such messages or, more precisely, who are vulnerable to becoming a threat to the British state. Therefore, identifying the nature of an individual's vulnerability to radicalisation is regarded as a crucial means of preventing terrorism (HM Government 2011).

Anxieties around the radicalisation of Muslim children and young people intensified during 2015 in the wake of reports of young British Muslims travelling to Syria to join the self-styled 'Islamic State' (Casciani 2015). In the same year, Parliament passed the Counter Terrorism and Security Act (CTSA 2015), which extended the role of schools and children's services to include 'protecting children from the risk of radicalisation' (DfE 2015b: 5). It also imposed a new legal duty (the 'Prevent Duty') for all public bodies, and individuals who work for them, to have 'due regard to the need to prevent people from being drawn into terrorism' (CTSA 2015, Section 26). This became effective 1 July 2015 and requires staff to be 'alert to signs of vulnerability to radicalisation' in a child's life, whether at home or elsewhere, and 'report them to the police'. The Prevent Duty is underpinned by statutory guidance to institutions (HM Government 2015), which requires them to promote British values to 'keep children safe' and ensure their welfare. In addition, staff must have a good understanding of PREVENT, be trained to recognise a child or young person's vulnerability to being drawn into terrorism, be aware of programmes that can deal with this issue and make appropriate referrals to 'CHANNEL', the government's de-radicalisation programme.

Originally piloted in 2007, the CHANNEL programme supports the PREVENT objective of protecting and supporting 'vulnerable' individuals who are exhibiting radicalised behaviour (HM Government 2011). It does so by working in partnership with local authorities, education, health, social services, youth services and communities to reduce or manage risk, and relies on referrals from teachers, social workers, youth workers, doctors, nurses, council workers etc. (HM Government 2012). The CTSA 2015 mandated the setting up of CHANNEL panels to ensure multi-agency risk assessment in line with child protection and safeguarding protocols. The role of the multi-agency panel is to develop appropriate 'support packages' for referred individuals who are 'channelled' away from extremism through interventions delivered by 'experts' in de-radicalisation. CHANNEL operations are shrouded in secrecy and specific information is not made public. However, figures released by the National Police Chiefs Council (following a freedom of information request from the Press Association) revealed that in the year 1 July 2015 to June 2016, of the total 4,311 people referred to CHANNEL, 2,311 were children and young people, an 83 per cent rise on the previous year. Of these, 352 were aged 9 years or under, 989 were aged between 10–14 years and 970 were aged 15–17 years (Press Association 2016).

The British government claims that PREVENT represents a non-partisan approach to the threat of radicalisation and is not targeted at a particular ethnic or religious group. This claim is difficult to sustain given that official PREVENT guidance explicitly associates the terrorist threat with Muslim faith and culture. For example, readers are alerted to warning signs of behaviour that may indicate potential evidence of radicalisation such as, 'support for the Islamic political system; a focus on scripture as an exclusive moral source; a conspiratorial mindset; seeing the West as a source of evil in the world and literalism in the reading of Muslim texts'. These are set alongside examples of 'persuasive extremist narratives', such as those that seek:

- To explain why I/my family/my community am/are experiencing disadvantage/suffering/lack of respect e.g. perceived persecution, inequality, oppression by a governing class, national or international politics

- To explain why the conventional family/school/community solutions do not provide answers to the core grievances e.g. 'the law does not protect us, my family is isolated from "real life" and does not know what it is like for young people'

(Department for Children Schools and Families (DCSF) 2008: 16)

This official narrative pathologises explanations offered by young British Muslims to describe the oppressive sociopolitical contexts impacting their lives – namely routine experiences of racism, Islamophobia, social exclusion and racially motivated violence – effectively depoliticising debate about these lived realities (Ahmed 2009). Moreover, it decontextualises an understanding of motivations for actual or potential political violence, diverting attention away from the civil and political rights implications of heightened state surveillance and securitisation practices that are undeniably targeted at Muslim young people with the intention of delegitimising and diffusing their political dissent. As Rights Watch UK observes, 'the Prevent strategy is leaving a generation of young Britons fearful of exercising their rights to freedom of expression and belief' (2016: 4). As a result, young British Muslims are increasingly self-censoring, as illustrated in the following quotation from a young student:

I must constantly be cautious about what I am Googling, what I write and publish on blogs, or whether or not I should retweet something about politics, religion or even humanitarian crises – anything which might possibly be misconstrued as 'extremism'.

(Yousuf Zubair, *Independent*, 6 March 2017)

The governance of the political identities of young British Muslims is further exemplified in the advice given to schools and childcare providers in exercising their safeguarding role under PREVENT (DfE 2015b) in which schools have a responsibility to build children and young people's resilience to radicalisation by promoting 'fundamental British values' and enabling them to challenge extremist views. This is to be achieved through:

providing a safe environment for debating controversial issues and helping them to understand how they can influence and participate in decision-making. [...] Citizenship [...] should equip pupils to explore political and social issues critically, to weigh evidence, to debate, and to make reasoned arguments [...] pupils learn about democracy, government and how laws are made and upheld. Pupils are also taught about the diverse national, regional, religious and ethnic identities in the United Kingdom and the need for mutual respect and understanding.

(DfE 2015b: 8)

A number of paradoxes are evident within this statement. First is the contradiction in the British state's apparent commitment to facilitating children and young people's political participation through an invitation to dialogue and critical debate on 'political and social issues', while simultaneously criminalising and closing down particular forms of political expression it deems inconsistent with 'fundamental British values'. Second, is the inculcation of young Muslims to unite around 'shared British values', where ' "British values" are held up as the "gold standard" for democratic citizenship and Muslim culture, traditions and values are pathologised as "other" and considered a threat to British society' (Coppock 2014: 122), leaving little room for 'mutual respect and understanding'. Third is the assumption that schools are a 'safe environment for debating controversial issues'. How is it possible for young Muslims to safely 'explore political and social issues critically' when, as described above, giving voice to their experiences of injustice and oppression is cited as an indicator of their potential radicalisation? In this sense, it is safer for Muslim young people to remain silent lest they run the risk of being referred to the CHANNEL programme. And so, paradoxically, within PREVENT, liberal-democratic 'citizenship' education becomes a vehicle for the political repression of young British Muslims, pointing towards a more sinister governance agenda, 'namely the shaping of a Muslim identity towards the formation of an idealised norm of the moderate Muslim' (Coppock 2014: 122).

In the above context, young British Muslims have faced a heightened state of hostility towards their faith and culture. Their existence has become increasingly securitised and all are deemed suspect and deserving of the state's gaze placed upon them. They do not have to be found to be committing any crimes; that they are Muslim is sufficient reason for them to be represented as dangerous and risky to the Western mind (Choudhury and Fenwick 2011), as witnessed in the following quotation from the same young British student cited above:

> On a day-to-day basis, I am hyperconscious about where I am sitting in a café or a park, when I do my daily Quran reading – who around me might see the Arabic writing on my laptop screen or mobile phone app and feel threatened or incensed? Might they even call the police or refer me to a Prevent channel, as has happened to others? If I was doing my French homework, I know I would hardly be noticed, but I worry terribly about the piercing eyes around me when reading Arabic, especially the Quran.
>
> (Yousuf Zubair, *Independent*, 6 March 2017)

The counterterrorism measures introduced through PREVENT and supported through the CHANNEL programme demand permanent monitoring and surveillance of the lives of young British Muslims, as witnessed in practices such as the insertion of police cameras in streets with a concentration of Muslim populations, or in the surveillance of schools, or increased stop-and-search practices of young Muslim men (Isakjee and Allen 2013). These practices are legitimised in the name of state security and serve to govern young Muslim people's potential political resistances, ensuring that they refrain from expressing or even affiliating with people and ideas that do not fit the hegemonic mould; to do so would be to invite intervention. Such measures have been criticised for their pathologising and alienating effects, for the increased intrusion of privacy and for the erosion of democratic rights and liberties for Muslim young people (Breen-Smyth 2014; Choudhury and Fenwick 2011; Coppock and McGovern 2014; Kundnani 2012; McGovern 2011; Ragazzi 2016; Rights Watch UK 2016).

On *becoming* radicalised: 'making up' the young radicalised Muslim

> Sometimes, our sciences create kinds of people that in a certain sense did not exist before. I call this 'making up people'.
>
> (Hacking 2006: 23)

The discourse of vulnerability to radicalisation within PREVENT rests principally upon the dual construction of the young British Muslim as both a politically risky subject (terrorist in the making) and a psychologically vulnerable subject (human becoming in need of protection). In this, the risk–vulnerability nexus acts as a powerful mechanism for the governance of Muslim young people's political identities through the normalisation and routinisation of surveillance and monitoring practices outlined above, justified as safeguarding their 'childhood' and 'youth' (DfE 2015a, 2015b). Here, we explore this risk–vulnerability nexus in more detail, drawing on theorisation from the sociology of childhood and Ian Hacking's (2006) critique of the role of 'psy' sciences in 'making up people' (23).

The tensions inherent in children and young people's political status in society are illustrated in the relationship between their protection rights and participation rights, as articulated in the United Nations Convention on the Rights of the Child 1989 (CRC). Despite the rhetorical emphasis on principles of autonomy and self-determination (derived principally in Article 12), the default position in practice leans towards protectionism, inhibiting the scope for political agency. Theorisation from the sociology of childhood reveals how this bias reflects institutionalised discourses of childhood vulnerability, where a distinction is made between the inherent vulnerability (a consequence of biological immaturity) and structural vulnerability of children, that 'comes about as a consequence of, and subsequently serves to reinforce, social and political mechanisms that reduce children's power, fail to take their agency into account and disregard their rights' (Powell and Smith 2009: 138). It also reflects an assumption that childhood innocence is undermined or threatened by politics and therefore children and young people should be excluded from the political world, *in their own best interests*.

These tensions are also rooted in the construction of children in the CRC as both future adult citizens and as citizens in their own right – that is, as both human beings and human becomings (Qvortrup 1994). In this, the 'psy' disciplines perform a crucial role in constructing, maintaining and giving scientific credibility to the Western, essentialised ideal childhood for the adult-in-waiting, and provide the state and its agents with technologies for correcting pathological, 'other' childhoods (Nieuwenhuys 1998; Rose 1985). Thus, 'the history of child welfare law, policy and practice is replete with examples of "interventions" aimed at restoring children to an imagined "normal" childhood, evidencing the consequences for those children who do not behave as "children"' (Coppock and McGovern 2014: 249). It is this history that continues to shape and inform contemporary state responses to young Muslims perceived as a potential threat to the British state, who 'are caught between processes of disciplinary normalisation, aimed at the production of governable bodies, and exercising their "rights and freedoms" as social and political actors' (Coppock 2014: 118).

The 'psy' concepts, technologies and practices evident in the PREVENT and CHANNEL programmes are illustrative of a process philosopher Ian Hacking (2006) describes as 'making up people' (23). He observes: 'creating new names and assessments and apparent truths is enough to create new "things". Making up people would be a special case of this phenomenon' (23). Following Hacking's analytical framework, we explore how the British state has 'made up' the young Muslim as 'vulnerable to radicalisation' through embedding particular forms of 'psy' discourse, knowledge and expertise in the institutional monitoring and surveillance practices of state professionals charged with protecting and safeguarding children and young people.

The classification: childhood radicalisation as 'another form of abuse'

With the passing of the CTSA 2015, childhood radicalisation became discursively framed as both a psychological disorder and another form of child abuse, effectively depoliticising the radicalisation process and embedding the securitisation agenda within existing protocols and processes for safeguarding vulnerable individuals (DfE 2015a, 2015b). In mobilising safeguarding discourse, the state distances itself from connotations of punitive intervention by giving the impression of a more palatable brand of normalisation – child saving (Parton 2014). Following Hacking (2006), we suggest that the classification of 'vulnerability to radicalisation' is neither an objective state, nor has it always been 'a way to be a person' (23). It is a socially constructed descriptive categorisation of persons, the widespread use of which emerged during the early years of the twenty-first century and which is applied disproportionately to Muslim young people who rapidly became a 'moving target' (23) for pseudo-scientific inquiry, institutional surveillance and sociopolitical control. Indeed, it is characteristic of the wider insidious trend towards the governance of populations through the psychologisation of social problems, whereby authoritarian state practices (reframed as benevolent rescuing interventions) 'move over time to become covert normalisation such that those who do not fit the norms are rendered deficient or pathological' (Burman 2012: 431). 'Making up' the classification of vulnerability to childhood radicalisation via the PREVENT and CHANNEL apparatus thus transforms youthful non-conformity and political dissent into a medico-legal 'problem' for which assessment, diagnosis and intervention is justified in the child or young person's best interests. The social control is therefore made to seem reasonable and legitimate, making it harder to resist. Meanwhile, Muslim children and young people are denied social and political agency through being both racialised and pathologised. In this sense, childhood radicalisation risk represents not so much a 'condition', but rather the manifestation of complex social, cultural and political factors implicated in the management of political dissent within Muslim communities and the creation of docile bodies.

The people and the institutions

Discursively constructed as vulnerable to radicalisation, Muslim children and young people constitute a 'moving target' (Hacking 2006: 23) for institutions and authorities charged with responsibility for protecting them (DfE 2015a). As established earlier, the Prevent Duty to have 'due regard to the need to prevent people from being drawn into terrorism' (CTSA 2015, Section 26) demands the permanent monitoring and surveillance of all aspects of the lives of Muslim children and youth, resulting in the full mobilisation of the child welfare infrastructure of the disciplinary state in the management of dissent and governance of young people's politics. In this, a climate of hyper-vigilance has developed amongst practitioners, to the extent that a child who misspelt the word 'cucumber' (misinterpreted by a teacher as 'cooker bomb') and a child who wrote about living in a 'terraced' house (misunderstood as a 'terrorist' house) raised alarm in school and resulted in referrals to Multi-Agency Protection Arrangements (Rights Watch UK 2016). This hyper-vigilance extends into the institution of the Muslim family where parents too are recruited by the state into governing their offspring's political allegiances. Parents are 'encouraged' to keep a watchful eye on their children's beliefs, attitudes, thoughts and activities, to monitor the clothes the children wear, the language and phrases they use and the websites they visit and report any suspect activity to the police – a measure many parents see as spying on their children (Awan and Guru 2016). Not only does this reflect the wider societal antagonism towards expressions of Muslim faith and culture, but also points to institutionally racist discourse that stereotypes Muslim parents as 'inadequate', unable to socialise their children and seeks to mobilise professional interventions to assist in 'appropriate' integration of dissenting Muslim youth into mainstream British politics and society. Furthermore, as we discuss later, it poses significant challenges for practitioners in maintaining social justice and human rights-informed practice with Muslim children, young people, families and communities.

The knowledge and the experts

Hacking (2006) highlights the pivotal role of knowledge (assumptions taught, refined and disseminated within institutions, seminars, conferences and training programmes) and experts (those who generate the knowledge, who judge if it counts as knowledge, and put it into practice) in the process of making up people. He describes 'the looping effect' (23) whereby the knowledge that the experts generate within institutions guarantees their status as experts who then study, advise or control those who are classified. This is evident in the PREVENT and CHANNEL programmes where a lucrative 'psy' knowledge industry has emerged around the psychology of terrorism, exemplified both in the commissioning of research in relation to screening and assessment tools to identify vulnerability to radicalisation and in the development and delivery of training programmes 'to assist individuals covered by the requirements of the Prevent Duty' (HM Government 2016: 4).

'Psy' knowledge has long been utilised by the state in the governance of its citizens (Burman 2012; Rose 1985). Positivistic 'psy' technologies (psychometric tests, diagnostic screening tools, profiling techniques etc.) superficially bear the hallmarks of objective science, conveying trustworthiness and the promise of truths and certainties in an uncertain, chaotic world. However, closer scrutiny of the knowledge base underpinning PREVENT and CHANNEL raises far-reaching questions regarding the legitimacy and uses of 'psy' research by the British state to bolster its strategic governance of the politics of young Muslims (CAGE 2016). For example, both the core introductory training programme for practitioners, Workshop to Raise Awareness of Prevent (commonly known as WRAP Training) and the statutory tool used by practitioners to assess childhood radicalisation risk, the Channel Vulnerability Assessment Framework (HM Government 2012), are based on a model known as 'Extremism Risk Guidance 22+' that posits 22 signs of vulnerability to radicalisation. Psychologists working for the National Offenders Management Service developed this model in secret. The original research study is classified (purportedly in the interests of national security) and has never been published. This means that it has not been subjected to scientific scrutiny or peer review, raising issues of reliability and validity. The researchers Lloyd and Dean have subsequently conceded that 'the manner in which it has been applied by the government goes beyond the parameters set' (CAGE 2016: 11) and that their failure to 'factor political grievance into the modelling [...] was "perhaps an omission"' (CAGE 2016: 4). This is crucial given that a central feature of the risk–vulnerability discourse in the Channel Vulnerability Assessment Framework is the reframing of normative expressions of political consciousness-raising as psychopathological traits associated with susceptibility to radicalisation, as illustrated in the following extract:

- May begin with a search for answers to questions about identity, faith and belonging

- May be driven by the desire for adventure and excitement
- May be driven by a desire to enhance the self esteem of the individual and promote their 'street cred'
- Is likely to involve identification with a charismatic individual and attraction to a group which can offer identity, social network and support.

(DCSF 2008: 17)

Furthermore, the framework lists a series of 'psychological hooks' which may lead a young person to be drawn into extremism. Again, perversely, the majority of these are self-evidently normative, such as: 'feelings of grievance and injustice; a need for identity, meaning and belonging; a desire for status; a desire for excitement and adventure; a desire for political or moral change; being at a transitional time of life' (HM Government 2012: 2). Read outside of this psychopathologising framing, these features may be reasonably understood as indicative of young people's *political* and *rational* engagement with the world around them, not as signifiers of a 'vulnerable' or 'disturbed' mind (Coppock and McGovern 2014). Nevertheless, despite fundamental methodological flaws, both WRAP Training and the Channel Vulnerability Assessment Framework are firmly established as mandatory tools of governance within children and young people's services, with clear implications in providing the intellectual justification and the technological know-how for governing the political identities of young British Muslims under the guise of protecting them.

In the service of the state? Troubling the social work role in the surveillance and governance of the politics of young British Muslims

The introduction of the Prevent Duty has produced an ethically fraught practice context for social workers in their engagement with young British Muslims and their families. Understandably and legitimately, children and young people want to ask questions about war in the Middle East, to explore ideas about Da-esh and the 'Islamic State', to voice their opinions about the international refugee crisis and to express anger, individually and collectively, at the UK's policy responses to these situations.

Should social workers collude in the governance of these children and young people's political identities by exercising strict adherence to PREVENT and CHANNEL guidance and the Prevent Duty, which requires that they report them to the counterterrorism authorities for 'de-radicalisation'? Or do they risk professional censure (or even individual prosecution) by supporting children and young people's rights to freedom of expression, thought, conscience and religion and association (CRC Articles 12, 13, 14 and 15)? And what about the long-term consequences for young British Muslims if social workers collude in the construction of risk identities? Police and security data will be held on children's services files indefinitely, potentially producing new sets of risk identities into their adult futures (Awan and Guru 2016; Bonino 2013; Choudhury and Fenwick 2011; Mythen *et al.* 2009). These questions pose moral and professional dilemmas that go to the heart of the value base of the profession. Such dilemmas are by no means new; social workers have always had to balance contradictory tensions in regulating the state/child/family nexus; exercising care and control, promoting rights and enforcing responsibilities on behalf of the state (Parton 2014). However, these tensions have intensified with the introduction of the Prevent Duty, which is clearly an instrument of governance *both* of social workers and of the children and young people with whom they work.

The history of children's social work demonstrates how in times of perceived societal crisis social work becomes increasingly risk averse and shifts sharply towards child rescue thinking (Parton 2014). At such times, the already precarious balance in the relationship between children, families and social work (as an established instrument of state governance) becomes even more contentious as formal statutory intervention invariably escalates. This is the current situation in the UK, where tensions in the role of social work and social workers in governing the political identities of young British Muslims on behalf of the state are acute.

Academics, practitioners and human rights activists are increasingly concerned about the intersection of child protection legislation with counterterrorism strategies (Awan and Guru 2016; Stanley and Guru, 2015). For example, there has been a sharp increase in the use of draconian wardship proceedings to prevent children and young people from travelling to Syria, on the grounds that they have been radicalised, or are at risk of radicalisation (Ashton 2016). Wardship is one of the most powerful legal tools in Britain, where the High Court exercises its inherent jurisdiction to secure legal guardianship over a child under 18 to ensure their safety and protection. The fact that social workers are resorting to its use reflects the pressure they are under to acquiesce to the state's securitisation agenda.

As is often evident in child protection practice (Parton 2014), this pressure is intensified by a media-fuelled 'moral panic' around young British Muslims attempting to travel to Syria to join the 'Islamic State'; reflected, for example, in sensational, racialised, trivialising reports of young Muslim women being 'seduced by the idea of becoming a Jihadi bride' (Casciani 2015). Passports have been seized and families ordered to comply with specific steps as dictated by a judge, including electronic tagging in some cases. In effect, the young people involved have been denied the right to political agency. Furthermore, they (and their families) have been denied involvement in decision-making, or in seeking practical solutions to the concerns held by social workers. This muscular form of social work practice is actively encouraged and legitimated by the appeal of persuasive individualised narratives of psychological disturbance and harm within official PREVENT and CHANNEL guidance, discussed above, which constructs and pathologises politically agentic Muslim children and young people as simultaneously 'at risk' and 'risky' (Heath-Kelly 2013).

In the current climate of heightened anxieties around preventing the radicalisation of young British Muslims, and the concomitant denigration and demonisation of their faith and culture, it is beholden on social workers to reflect critically on their contradictory position both as allies of oppressed Muslim children, young people and communities and as potential 'agents of the state'. In this, there needs to be a *radical* response from the social work profession, akin to that of their teaching counterparts who have voted overwhelmingly to reject PREVENT and have called on the government to involve the profession in the development of alternative approaches to engaging young people in political debate (Adams 2016).

The challenge of maintaining social justice and human rights-informed practice has never been greater, or more difficult, but it is our view that a re-imagining of the radical social work tradition of the 1970s (Bailey and Brake 1975) could offer the scope for alternative, rights-informed approaches to theory and practice to emerge, such as collective resistance to hegemonic individualising theories of psychological dysfunction and the development of constructive models of assessing risk through open dialogue with young people. Rather than uncritically accepting vulnerability to radicalisation discourse in the name of the state's duty to protect, efforts should be made to understand and respect young people's political perspectives and choices and to seek open conversations with them about potential risks. It is profoundly important that social workers do not to dismiss Muslim young people's views and actions as irrational and deviant, but rather understand and respect them as demonstrations of political agency – however misguided, unpalatable or frightening as their political views may appear to the adults around them, or indeed to the state.

Conclusion

In this chapter, we critically analysed the British government's PREVENT counterterrorism and anti-radicalisation strategy, both as a rationale and vehicle for the pre-emptive monitoring, surveillance and governance of the politics of young British Muslims. We problematised the embedding of PREVENT within child protection and safeguarding protocols through the process of 'making up' the official classification of 'vulnerability to radicalisation', in which young British Muslims are pathologised as both 'at risk' and 'risky' to the state and therefore in need of 'saving' *in their own best interests*. Moreover, we identified how this new discursive construction derives from the same dominant forms of 'psy' knowledge production and associated disciplinary networks and practices that regulate and govern the spaces of childhood and youth generally. We argued that these processes and practices obstruct and delimit the political agency of young British Muslims in the interests of the state. Finally, we have discussed how the introduction of the legal duty to assess and report childhood radicalisation risk poses serious challenges for public-sector professionals expected to comply with it.

We illustrated how, through the PREVENT strategy and its associated CHANNEL de-radicalisation programme, the British state is driving social work in ways that are inconsistent with the core values of the profession, producing muscular forms of practice that assist the state in its aim to repress expressions of youthful political dissent. We conclude that it is imperative that social workers resist the individualising, pathologising presumptions and repressive practices entailed in PREVENT and reclaim the radical social work principles of equality and social justice that symbolise what it means to do *radical* social work.

References

Adams, R. (2016). 'Teachers back motion calling for Prevent strategy to be scrapped'. *Guardian*, 28 March. Available from www.theguardian.com/politics/2016/mar/28/teachers-nut-back-motion-calling-prevent-strategy-radicalisation-scrapped [accessed 9 February 2017].

Ahmed, S. (2009). *Seen and not heard: voices of young British Muslims*. Markfield, Leicestershire: Policy Research Centre.

Ashton, J. (2016). Radicalisation of children: wardship – an old measure to tackle a very new problem. *Student Law Review*, 78: 33–34.

Awan, I. and Guru, S. (2016). Parents of foreign 'terrorist' fighters in Syria – will they report their young? *Ethnic and Racial Studies*, 40(1): 24–42.

Bailey, R. and Brake, M. (eds). (1975). *Radical social work*. London: Edward Arnold.

Bonino, S. (2013). Preventing Muslimness in Britain: The normalisation measures to combat terrorism. *Journal of Muslim Minority Affairs*, 33(3): 385–400.

Breen-Smyth, M. (2014). Theorising the 'suspect community': counterterrorism, security practices and the public imagination. *Critical Studies on Terrorism*, 7(2): 223–240.

Burman, E. (2012). Deconstructing neoliberal childhood: towards a feminist antipsychological approach. *Childhood*, 19: 423–438.

CAGE. (2016). *The 'science' of pre-crime. The secret 'radicalisation' study underpinning Prevent*. London: CAGE Advocacy UK Ltd.

Casciani, D. (2015). 'The medieval law taking on Syria travellers'. BBC News, 31 December. Available from www.bbc.co.uk/news/uk-35199266 [accessed 9 May 2017].

Choudhury, T. and Fenwick, H. (2011). *The impact of counter-terrorism measures on Muslim communities*. London: Equality and Human Rights Commission.

Coppock, V. (2014). 'Can you spot a terrorist in your classroom?' Problematising the recruitment of schools to the 'war on terror' in the United Kingdom. *Global Studies of Childhood*, 4(2): 115–125.

Coppock, V. and McGovern, M. (2014). 'Dangerous minds'? Deconstructing counterterrorism discourse, radicalisation and the 'psychological vulnerability' of Muslim children and young people in Britain. *Children & Society*, 28: 242–256.

CTSA. (2015). *Counter Terrorism and Security Act 2015*. Section 26. London: The Stationery Office.

Department for Children, Schools and Families. (DCSF). (2008). *Learning together to be safe: A toolkit to help schools contribute to the prevention of violent extremism.* Nottingham: DCSF.

Department for Education. (2015a). *Working together to safeguard children: A guide to inter-agency working to safeguard and promote the welfare of children.* London: Department for Education.

Department for Education. (2015b). *The Prevent duty. Departmental advice for schools and childcare providers.* London: Department for Education.

Hacking, I. (2006). Making up people: clinical classifications. *London Review of Books,* 17 August 2006, 28(16): 23–26.

Heath-Kelly, C. (2013). Counter-terrorism and the counterfactual: producing the 'radicalisation' discourse and the UK PREVENT strategy. *The British Journal of Politics and International Relations,* 15(3): 394–415.

HM Government. (2011). *Prevent strategy.* London: The Stationery Office.

HM Government. (2012). *Channel: Vulnerability Assessment Framework.* London: The Stationery Office.

HM Government. (2015). *Prevent duty guidance: for England and Wales. Guidance for specified authorities in England and Wales on the duty in the Counter-Terrorism and Security Act 2015 to have due regard to the need to prevent people from being drawn into terrorism.* London: The Stationery Office.

HM Government. (2016). *Prevent: Training catalogue.* London: The Stationery Office. Available from www.gov.uk/government/uploads/system/uploads/attachment_data/file/503973/Prevent_Training_catalogue_-_March_2016.pdf [accessed 24 April 2017].

Isakjee, A. and Allen, C. (2013). 'A catastrophic lack of inquisitiveness': A critical study of the impact and narrative of the Project Champion surveillance project in Birmingham. *Ethnicities,* 13(6): 751–770.

Kundnani, A. (2012). Radicalisation: the journey of a concept. *Race and Class,* 54(2): 3–25.

McGovern, M. (2011). The dilemma of democracy collusion and the state of exception. *Studies in Social Justice,* 5(2): 213–230.

Mythen, G., Walklate, S. and Khan, F. (2009). I'm a Muslim, but I'm not a terrorist: victimisation, risky identities and the performance of safety. *The British Journal of Criminology,* 49(6): 736–754.

Nieuwenhuys, O. (1998). Global childhood and the politics of contempt. *Alternatives: Global, Local, Political,* 23: 267–290.

Parton, N. (2014). *The politics of child protection. Contemporary developments and future directions.* Basingstoke: Palgrave Macmillan.

Powell, M. and Smith, A. (2009). Children's participation rights in research. *Childhood,* 16: 124–142.

Press Association. (2016). 'Jump in CHANNEL referrals after duty to combat extremism introduced'. *AOL News,* 12 September. Available from www.aol.co.uk/news/2016/09/11/jump-in-channel-referrals-after-duty-to-combat-extremism-introduced/ [accessed 9 May 2017].

Qvortrup, J. (1994). Childhood matters: An introduction. In: J. Qvortrup, M. Bardy, G. Sgritta and H. Wintersberger (eds.), *Childhood matters: social theory, practice and politics.* Aldershot: Avebury Press.

Ragazzi, F. (2016). Suspect community or suspect category? The impact of counterterrorism as 'policed multiculturalism'. *Journal of Ethnic and Migration Studies,* 42(5): 724–741.

Rights Watch UK. (2016). *Preventing Education? Human Rights and UK Counter-Terrorism Policy in Schools.* Available from www.rwuk.org/research/ [accessed 13 July 2016].

Rose, N. (1985). *The psychology complex: psychology, politics and society in England 1869–1939*. London: Routledge and Kegan Paul.

Stanley, T. and Guru, S. (2015). Childhood radicalisation risk: an emerging practice issue. *Practice: Social Work in Action*, 27(5): 353–366.

Zubair, Y. (2017). 'As a young British Muslim, I've been kicked out of a mosque and told to stay away from Google – why do I feel like a terrorist?' *The Independent, Voices*, 6 March. Available from www.independent.co.uk/voices/muslim-young-british-islam-fundamentalist-terrorism-isis-mosque-religion-why-feel-like-one-a7613491.html [accessed 9 May 2017].

Chapter 9

Active citizenship and governmentality

The politics and resistance of young Muslims in the security state

Anisa Mustafa

Introduction

At the most basic level, citizenship describes 'a territorially bounded' political community where 'national belonging' is the 'legitimate basis of membership' (Soysal 2012b: 384). Since citizenship emerged with the rise of nation-states, it is associated with the culture, traditions and history of dominant groups that can sometimes marginalise minorities by asserting their own norms (Meer 2010). This chapter highlights the ways in which young British Muslims are being disadvantaged in the war on terror by a reformulation of the meaning of citizenship in the wake of resurgent nationalism (McGhee 2008) and securitisation (McGhee 2010). In addition to being subjected to intense surveillance and scrutiny post 9/11 and 7/7, young Muslims are facing demands to prove their loyalty to Britain by supporting such repressive measures (Brown 2010, Mythen et al. 2009; O'Loughlin and Gillespie 2012).

This chapter focuses on the politics of young British Muslim activists, who struggle against their marginalisation and demonisation as a consequence of the securitisation of citizenship. A polysemous term, securitisation, refers to processes through which security becomes the dominant function of the state. It is a condition in which exceptional state powers are appropriated in response to prospective rather than imminent threats to the public (Hussain and Bagguley 2012; McGhee 2010). The chapter elucidates how participants resist the increasingly strident and exclusionary term 'active citizenship' (McGhee 2008; Soysal 2012a) that signifies the cultural dimension of securitisation. Viewed through the interpretive lens of Michel Foucault's (1994a: 219–222) concept of 'governmentality', the term 'active citizenship' can be seen as a technology of power that is productive as well as repressive. Foucault's (1994a, 1994b, 1995) theory of governance argues that modern states rely less on direct force and more on surveillance to probe, manage and control the population. According to Foucault, expert knowledge is a technology of power used by states to shape the actions and choices of individual citizens. The effects of such power can be detected in the normalisation of ideas and practices, like 'active citizenship', that can have repressive consequences.

The following discussion is based on an ethnographic study exploring the political and civic activism of young British Muslims. The research aimed to understand how young Muslims responded to the often coercive effects of new ways of framing citizenship, particularly the notion of 'active citizenship' as it is invoked in the war on terror. The research was conducted in the Midlands, UK between 2012 and 2014, with young adult Muslims who were active in civil society. The focus of the research was on non-electoral or 'subpolitical' (Beck 1997) activities, typically associated with social movements. This chapter is based on interviews with 34 male and female participants, aged 17 to 37 years, engaging in a range of activities from anti-war and pro-Palestine campaigning to cultural/religious education and voluntary/charity work. Although the age range of the participants exceeded the traditional youth category of 18–25 years, this was a deliberate strategy to chart the experiences of young adults in the decade following 9/11.

To begin with, Foucault's concept of governmentality is explained in more depth, followed by a discussion on how this informs a reading of 'active citizenship' in public policy as a mechanism to control Muslim politics. Next, the chapter explores how young Muslim activists defy dominant notions of citizenship in complex and contradictory ways, exemplifying 'governmentality' through the forms of resistance it engenders. The conclusion surmises that cultural contestation is a less risky strategy to express political agency in a climate of securitisation, where increasingly repressive measures are being used to safeguard public security.

Governmentality, surveillance and knowledge

Foucault claims that power is 'pluralist; it is exercised from innumerable points, rather than from a single political centre' (Nash 2010: 21). Foucault (1994a, 1994b, 1995) argues that power in contemporary governance is exercised everywhere through a nexus of surveillance, knowledge and discourse. According to Foucault (1994a), this type of power emerged in the eighteenth century with a shift in modes of governance. As populations grew and became more complex, the use of direct force became a less efficient way to rule over territories. Governance moved towards the use of administrative methods of management based on gathering facts about individuals in society, in order to govern their 'conduct' (Foucault 1994b: 341). In this mode of government, power is less concerned with 'sovereignty' and more so with the 'welfare of the population' which it achieves 'through techniques that will make possible, without the full awareness of the people' the fulfilment of these ends (Foucault 1994a: 217). This is what Foucault (1994a: 219) has referred to as 'governmentality' – a rationality of government that is both 'an individualizing and a totalizing form of power' (Foucault 1994b: 332). It is so because it acts upon the choices and actions of individuals for the benefit of the whole. The production of knowledge is essential to this form of government in liberal states, since the people to be governed are 'endowed with a capacity for autonomous, self-directing activity' (Hindess 2012: 39). Governmentality is political rule that directs the autonomous individual's capacity for self-control towards achieving the goals of the state.

Foucault (1994c) identifies a significant modification in the way knowledge was construed around the time that 'governmentality' emerged. From being an act of inquiry, knowledge transformed into something that is 'organised around the norm, in terms of what was normal or not, correct or not, in terms of what one must do or not do' (Foucault 1994c: 59). The ascendency of the sciences, particularly social sciences, and professional expertise went 'hand in hand with the installation of new mechanisms of power' (Foucault 1988: 106). Empirically, Foucault applied his theory of power and knowledge to areas of social life such as mental illness, sexuality and crime. Using historical analysis, Foucault described how expert knowledge on these subjects was used to control socially undesirable or risky behaviours. The advent of official records on births, deaths, crimes, marriages and health in the population was not a purely benign and benevolent exercise, it also entailed surveillance through which an individual's choices could have punitive consequences. Foucault (1995: 209) refers to this transformation in the way that power was exercised as encompassing 'the formation of what might be called in general the disciplinary society'. This increasing professionalisation of knowledge, encroaching on many private aspects of life, is seen by Foucault as a tactic that inscribes desirable dispositions within citizens and produces subjectivities acquiescent to power. Governmentality as the 'conduct of conducts' (Foucault 1994b: 341) was not simply a productive exercise, it was also a repressive one since 'To govern, in this sense, is to structure the possible field of action of others'.

Knowledge that is deployed in the exercise of power 'not only assumes the authority of the "truth" but has the power to *make itself true*' (Hall 2010: 76, original emphasis) through discourse that 'transmits and produces power' (Foucault 1978: 101). Discourse in the Foucauldian sense signifies the use of language as a social practice that constructs objects in the world (Foucault 1981; Hall 2010).

Forms of power that operate through knowledge and discourse are hard to identify and challenge because they function as invisible and neutral forces. However, these forms of power can be studied by the effects they produce, through 'a way that is more empirical' which 'consists in taking the forms of resistance against different forms of power as a starting point' (Foucault 1994b: 329). Foucault (1994b: 329) suggests that 'Rather than analysing power from the point of view of its internal rationality, it consists of analysing power relations through the antagonism of strategies.' Social movements provide an apposite example as they challenge governmentality itself:

> They are an opposition to the effects of power linked with knowledge, competence, and qualification – struggles against the privileges of knowledge.
>
> (Foucault 1994b: 330)

Foucault's theory of power and knowledge expounds the voluminous literature that has proliferated after 9/11, much of which focuses on the assumed pathologies of Muslim citizens (Choudhury 2007; DCLG 2010; Home Office 2005; Wiktorowicz 2005). McGhee (2010: 24) argues that counterterrorism approaches have relied upon 'knowledge systems' derived from modern surveillance that allows the state to use 'strategies of governing from a distance'. Accordingly, counterterrorism has spawned a whole 'government-funded industry' of experts who claim their knowledge enables them to 'propose interventions in Muslim communities to prevent extremism' (Kundnani 2012: 3). More importantly, 'governmentality' explains how Muslim citizenship has become securitised through this approach that frames terrorism as a cultural problem within Islam. This can be observed in the way that the concept of 'active citizenship' has come to signify a burden of greater demands on citizens (Soysal 2012a), particularly young Muslims. As public security becomes central to the state's functions, active citizens are being redefined as supporters of counterterrorism policy, who are agents as well as subjects of surveillance (McGhee 2010). This makes the concept an instrument of power that limits the scope of British Muslim politics.

'Active citizenship' and securitisation

This section brings into focus the repressive effects that an emphasis on 'active citizenship' in government policy has had on Muslim citizenship. The emergence of 'active citizenship' as a central theme in governance corresponds with the incessant promotion of 'shared British values' as a 'solution to many of our woes from counterterrorism, to the lack of integration and community cohesion in the UK' (McGhee 2010: 128). In other words, 'active citizenship' has been promoted both through resurgent nationalism and securitisation, as the following discussion illustrates.

While globalisation has opened up possibilities for new modes of postnational citizenship (Sassen 2002; Soysal 2012a, 2012b) many Western states have responded to the challenges of growing diversity with resurgent forms of nationalism. This is often driven by moral panics over the assumed failure of migrants to assimilate, as expressed in a 'backlash against multiculturalism' (Vertovec and Wessendorf 2010: 6) after 9/11. This is significant for young Muslims because 'a weak sense of British citizenship' (McGhee 2008: 83) is seen as the cause of social conflict and a threat to national security (Modood 2007; Parekh 2008). Such fears are exacerbated by the growing salience of religious identification amongst younger Muslims (Ansari 2009; Cesari 2009), that is equated with rejection of British identity and interpreted as a risk factor for radicalisation (Kundnani 2009). In response, British policies on integration have emphasised a more vigorous and 'active' notion of British citizenship by privileging 'shared identity' and rejecting 'an unbridled multiculturalism' (Blunkett 2002 in McGhee 2008: 88). The way in which 'active citizenship' has been promoted by emphasising the need for 'loyalty, duty and responsibilities' (McGhee 2008: 43) implies a Muslim deficit in this regard. This sentiment resonates in the former prime minister of the United Kingdom David Cameron's assertion that:

> We have failed to provide a vision of society to which they [Muslims] feel they want to belong. We have even tolerated these segregated communities behaving in ways that run counter to our values.
>
> (BBC News 2011)

The suggestion that Muslims choose to live in segregated communities to pursue values that are 'counter' to British values places Islam in direct conflict with state-sponsored nationalism. This explains why studies in which Muslims prioritise their religious identity over their national identity (Ameli 2002; Choudhury 2007) are met with alarm and panic (Parekh 2008). A public policy focus on adoption of British values as a solution to poor Muslim integration has been criticised for ignoring the structural and economic barriers to inclusion (McGhee 2008, 2010; Parekh 2008).

Similar tactics appear in the deployment of 'active citizenship' in relation to security, where promoting 'fundamental British values' (Thomas 2016: 184) has become essential to counterterrorism since 9/11. As a result, 'public security' has become the 'ultimate value' in a 'new politics of vigilance' (Amoore 2007 in McGhee 2010: 13). Paradoxically, the concept of 'active citizenship' in this context inhibits political activism by young Muslims because it is seen as 'subversive' (Briggs 2010: 272). Spalek (2010: 792) argues that:

> Muslims' responsibilities as active citizens are being increasingly framed by anti-terror measures, which encourage internal community surveillance so that the responsible Muslim citizen is expected to work with the authorities to help reduce the risk of terrorism.

This suggests that Muslims are expected to demonstrate 'active citizenship' by supporting the state's counterterrorism measures. An example is the 'PREVENT' policy, introduced in 2007 as part of the multi-pronged CONTEST approach to counterterrorism (Brown 2010; Vertigans 2010). Embedded in youth work and community projects to build resilience against extremism, PREVENT has been criticised for deploying surveillance rather than education and political participation as strategies to address radicalisation (Thomas 2016). It has been described as a 'disproportionate reaction of repression, restrictions on civil liberties and scapegoating of specific communities' (Thomas 2016: 172). Through PREVENT, the government attempted to win the 'hearts and minds' of young Muslims (Brown 2010: 172) and propagate more 'moderate' interpretations of Islam. This was based on the belief, held by many national security experts, that the root causes of terrorism were firmly fixed in theological convictions

(Kundnani 2012). Within Muslim civil society, PREVENT created a climate where 'to make radical criticisms of the government is to risk losing funding and facing isolation as an "extremist", while those organisations which support the government are rewarded' (Kundnani 2009: 6). This created a pernicious distinction between good and bad Muslims. Moderate or good Muslims were those who supported government policies by agreeing to police their communities, while those who refused were deemed to be dangerous or 'bad Muslims' (Mamdani 2002: 766). The way in which good citizenship became contingent upon compliance with government policy represents the securitisation of citizenship (Hussain and Bagguley 2012; McGhee 2010; Thomas 2016).

Securitisation signifies a condition where anticipated or imagined risks are used to justify exceptional and 'pre-emptive' state powers that 'erode or eliminate traditional principles, standards and procedures of law' (McGhee 2010: 63). In other words, securitisation occurs when 'existential threats' allow the state to break free from normal, democratic rules (McGhee 2010: 42). This has been the case in Britain where deeply cherished and hard-won civil liberties, such as protection from detention without charge, have been compromised to ensure public safety. It is argued that greater police powers, such as increased stop and search and intrusive surveillance methods, are being used to suppress political dissent (Brown and Saeed 2014; Frost 2008). As such exceptional measures require public support, the state has been engaging in a 'politics of fear' (McGhee 2010: 2) that casts terrorism as 'a problem of Islamic practice and community life' (Thomas 2016: 174). The concept of governmentality explains how measures like PREVENT have been made possible by the negative portrayal of Islam as a detriment to national security. As Brown (2010: 181) argues, 'The ability to represent British Muslims as subversive, dangerous and threatening therefore becomes pivotal to the contemporary articulation of security in the UK context'. According to Vertigans (2010), the threat to civil liberties posed by the war on terror is tolerated because Muslims are seen to be the target of its repressive policies.

As a consequence of securitisation, 'young Muslims inhabit a contradictory and ambiguous space in relation to their values and identities, being depicted as a high-risk group while simultaneously being exhorted to assimilate more fully into British society' (Mythen *et al.* 2009: 740). The framing of Muslim citizenship through the lens of security manifests the workings of governmentality, as the state's powers are exercised through hegemonic concepts like 'active citizenship'.

Resistance to 'active citizenship'

Two distinct forms of civic consciousness or models of activism were identified in this study. One marked by overt and explicit resistance to 'active citizenship', as defined by the state, while the other approach demonstrated greater compliance but with 'implicit' resistance. The contrasting ways in which young

British Muslim activists described citizenship demonstrates how the hegemonic framing of national identity is resisted but also how it permeates and shapes political subjectivities in complex ways.

Counter-hegemony and explicit resistance

Given extant literature on the alienation and disaffection many Muslims feel towards their national identity (Ameli 2002; Choudhury 2007; Parekh 2008), it was not surprising to find a number of participants expressing negative views on 'British' citizenship. The following quote from a pro-Palestine activist highlights how British identity, when associated with the state, creates inner conflict for those who are uncomfortable with Britain's colonial past. As Haseena explained:

> I just feel like I'm not a British citizen because I know what that stands for, I know what that symbolises, that's you know all the bad things about capitalism, white supremacy, patriarchy so I don't want anything to do with that.
>
> (Haseena[1], 21, female, student activist)

This highlights why British national identity troubles some participants, as it is seen as endorsement of the nation's imperial history. Haseena associates Britain's colonial tradition with oppressive ideologies, like capitalism and patriarchy, which she believes are inextricably woven into the concept of British citizenship. As Haseena implies, this makes it difficult for her to assume this identity. Haseena's focus on state ideologies and practices signifies an understanding of citizenship as a 'vertical' relationship between the citizen and the state, in contrast to 'horizontal' relationships with other citizens (Thomas 2016: 183). It is the latter aspect of citizenship that has received greater attention in government efforts at civic renewal since 2001 (Thomas 2016). When framed in terms of bonds with other citizens, for many participants, their British citizenship was challenging due to experiences of racism. This was the case for Romana, who grew up feeling excluded from British society due to perceived prejudices against Muslims.

> I'm not really a citizen of like anything ... I don't feel included in this country ... you're always like ... I mean, especially as a Muslim, like, you're just like, people don't want you here, you take their jobs ... you like murder their parents or whatever.
>
> (Romana, 20, female, student activist)

In contrast to Haseena, who rejected national identity because of its symbolic connotations, Romana was describing her experience of rejection from British society because of her Muslim identity. This created an obstacle to identifying as a 'citizen'. She was not alone in this, as other interviewees expressed similar experiences of unfair and discriminatory treatment that made it difficult to feel a sense of belonging to Britain. As Parekh (2008) has explained, in a context of such hostility many young Muslims turn to their faith because it offers a sense of belonging through a more positive identity. However, faith identities can also be risky for young Muslims because of the real or imagined compulsion to give priority to national identification post 9/11, as Amna elaborates:

> When you see your mother walking down the road getting spat at or you know got called a Paki by little kids because that's the culture that they have, where they treat us as different, as outsiders, stuck us into our little areas … then how can you identify with that, how can you say that I'm British, it would be a bit of a cop out to be honest. I'd feel untrue to myself if I did that. But I can see why they [government] find it worrying but what are they gonna do? Keep sending in PREVENT people and trying to tell young kids that they shouldn't identify with Muslims across the world and they should feel more British? It's not going to work.
>
> (Amna, 31, female, student activist)

Amna's critical account emphasises how securitisation can further complicate the difficult relationship that many young Muslims have with their national identity. While racism in the everyday experience of citizenship creates barriers to inclusion, seeking alternative sources of esteem in the universal embrace of religion carries risks of punitive consequences. This is because lack of identification with Britain, along with other factors such as heightened religiosity, can warrant a referral to counter-radicalisation programmes under PREVENT (Thomas 2016). Amna's comments demonstrate how the idea of 'citizenship' can become repressive when it is associated with policies to tackle terrorism.

These critical comments about national identity are interpreted as being 'explicit resistance' to demands being made on Muslims to conform to hegemonic definitions of citizenship. However, there were more 'implicit' ways of contesting hegemony. The following sections reveal these by focusing on attitudes towards citizenship that appeared to be more complicit with statist discourses.

Compliance or implicit resistance?

Not all the participants were contemptuous of nationalism, indeed, some expressed ideas that contained traces of hegemonic discourses on 'active citizenship'. Such views were more common amongst participants whose activism was of a less dissenting nature, such as charitable work or cultural and religious education. These views contrasted with the previously discussed notions of citizenship as explicit resistance, even if the affirmation of British identity was not unqualified.

> My identity itself is a British Muslim, I'm not British, I'm not just a Muslim,
> I'm not just Yemeni, I'm a British Muslim that's how I think of it ...
> because I feel like that in itself is a culture.
>
> (Arya, 19, female, student/community activist)

While British identity carried some sense of pride for Arya, it was also tempered by the equal weight given to religious identification. For Arya, being a 'British' Muslim was a valued feature of her identity since she was able to distinguish this from other ways of being a Muslim in countries like Yemen. Identifying as British was important in her bid to counter the negative perception of Muslims in Britain and promote a better public understanding of Islam. Like Arya, many other participants who were less critical of the state tried to combat Islamophobia by projecting positive examples of Muslim citizenship. These efforts could sometimes appear to support government and media discourses blaming radicalisation on the failure of Muslims to integrate (Kundnani 2009; Mythen *et al.* 2009).

> If we throw ourselves out and get involved in all the community work that
> is out there, it could be something simple as coffee mornings or street clean-
> ups or going to visit sick in hospital or whatever, and that will change
> people's perception of Islam. 'Cause at the moment we're seen as a very
> restricted, very closed community.
>
> (Abid, 25, male, community activist)

Abid's belief that challenging Islamophobia requires a change in the behaviours of Muslims reflects a similar logic to the assumptions underpinning British integration and counterterrorism policies discussed earlier. It has been argued that counterterrorism strategies targeted indiscriminately at Muslim communities, rather than criminal suspects, is fuelling Islamophobia (Frost 2008; Mythen *et al.* 2009), yet Abid emphasises Muslim culpability for negative perceptions of Islam. This highlights how dominant discourses can be internalised in ways that obscure their roots in state technologies of power (Marinetto 2003; Soysal 2012a). This is particularly evident where participants echo discourses framing Muslims as a 'cultural and political threat' (Parekh 2008: 99).

> I think they [Muslims] have so much to be grateful for in this country;
> imagine if the white people were in Pakistan and if they had done half the
> things that Muslims are doing in this country, you know, how tolerating
> would Pakistan be of that.
>
> (Beena, 21, female, student/community activist)

In this quote, Beena is echoing the view, held by many critics of multiculturalism, that Muslim demands for cultural and religious accommodation are incompatible with the secular, liberal values of the West (Meer 2010). Despite appearing otherwise, Beena's comments were driven by a desire to combat Islamophobia. This was evident from her political goal to promote better education and employment for Muslim women, to challenge a common public impression that Islam undermines gender equality. However, this also made her more critical of what she believed to be conservative cultural practices amongst some Muslims.

This section has explored how young Muslim activists sometimes reproduced hegemonic discourses. As the next section argues, these accounts do not lack political agency but rather they signify resistance that is commensurate with forms of power described as 'governmentality'.

Acts of resistance and governmentality

Despite being reluctant to fully embrace nationalist conceptions of 'active citizenship', most participants did express a strong sense of civic duty compelling them to be 'active' in British civil society. The young Muslim activists in this study described and embodied notions of active citizenship premised on an obligation to serve 'humanity' or the 'community'. Far from being a sign of subversion or separatism (Briggs 2010), as security-minded politicians fear, such challenges to the limits of citizenship have been a typical feature of social movements (Nash 2010). Often invoking human rights or global citizenship as the basis for equality and inclusion, social movements have 'significantly blurred the conventional dichotomy between national citizens and aliens' (Soysal 2012b: 385). Dania explains:

> [...] to me, citizenship is a very wide and vast thing, like, I don't strictly see myself as a citizen of this country or as a citizen in as far as [...] my rights and responsibilities, it's not honed only to the UK or to Britain and it's not to Palestine either, like I see myself as having a social responsibility for the whole of the world, like a global citizen.
>
> (Dania, 23, female, student/community activist)

Dania's affinity for global citizenship chimes with the growing appeal of 'postnational citizenship' amongst a host of actors, including environmentalists and migrants (Soysal 2012b). The concept denotes the 'experiences and practices associated with citizenship' that 'exceed the boundaries of the territorial nationstate' (Sassen 2002: 278). As previous research has revealed, for many young Muslims 'the unitary world of the nation state' does not square easily with their 'plural political imaginaries' (Back *et al.* 2009: 4). Participants in this study exemplified such imaginaries by expressing multiple motivations for active citizenship, all drawing upon universal ideals transcending nationalism. Zahir illustrates how religious faith informs his account of citizenship:

> Sincerity is a very basic, fundamental belief of Muslims. You do things not for anybody else; you're doing it to please your Lord and God. So in a sense

of citizenship, it should be why you do things, how sincere are you in what you do. It could be helping an old woman off the bus or helping somebody with shopping bags, whatever it is you're doing it's not for the sake of anybody, you're doing it to please, if you're spiritual, your Lord and if not, if you do not believe in God, you're doing it because it's the right thing to do.

(Zahir, 25, male, community activist)

Such normative ideas and logics disturb nationalist ideals of citizenship through cultural contestation. By 'subverting' the given meanings of cultural practices and redeploying them for their 'own interests and agendas' these young activists engaged in a 'recoding' of social norms that signifies agency (Mahmood 2005: 6). Many participants who were 'explicit' in their resistance would dismiss 'implicit' forms of resistance for posing no real challenge to power holders. Nevertheless, the different values and beliefs guiding these understandings of citizenship can be seen as subversive of nationalist ideals. They also bring into critical focus the 'securitisation of Muslim subjects' (Mavelli 2013: 159) in the war on terror. As Mahmood (2005: 9) observes, there is a danger of assuming that 'resistance' has a universal meaning that may exclude acts that are not perceptibly defiant. Whether or not participants in this study agree with the interpretation of their actions and views as signs of resistance, Mahmood's (2005: 33) insistence on sensitivity to 'historically contingent arrangements of power' warrants consideration of the disproportionate impact of securitisation on their political subjectivities. The following quote is indicative of the constraints faced by activists:

I just see it [citizenship] as more of a rebellion because to be honest with you, what is a good citizen? I mean in their eyes, you know, you just sit down, don't say anything, don't question anything [...] yeah, just let it be and we'll sort it out eventually [...] I mean you try to be a good sort of citizen, like, you tried your best but then again, you're still discriminated against. Like, what are the rules and regulations? Like, what is a step too far for being an active citizen, if you're understanding what I'm saying? Like, again, all the rights to protest, to speak out, freedom of expression, freedom of speech is all being sort of [...] what can I say? [...] They're starting to fit it into sort of like, little boxes, like [...] and if you step outside of that, you know, you're done for, basically.

(Mona, 27, female, photographer and blogger)

This implies an acute awareness of the limits on political action under conditions of securitisation. What Mona describes as 'little boxes' that 'active citizenship' is expected to 'fit' into designates for her the 'possible field of action' (Foucault 1994b: 341) for politics. As Foucault (1995) argued, surveillance as a form of discipline subdues the body but also bounds the imagination and speech of those under scrutiny. The cautious manner in which Mona mentioned 'the step too far' for active citizenship, seeking unspoken 'understanding', suggests a fear of harmful repercussions. This is clear from her belief that if you do 'step outside' of the boxes you are 'done for'.

Mona's observations indicate the limits of securitised citizenship and explain why many participants in the study were reticent on the subject of security or 9/11, rarely speaking of either voluntarily. The silences and elisions around the war on terror were striking in the interviews and revealing of the repressive effects of PREVENT. The concept of governmentality helps to elucidate why discursive challenges to given meanings of citizenship are a necessary and appropriate strategy in contesting securitisation. These forms of resistance reveal the salience of knowledge and discourse as contemporary modes of social control. As Foucault (1978: 95) argued:

> Where there is power, there is resistance, and yet, or rather, consequently, this resistance is never in a position of exteriority in relation to power.

This implies that cultural contestation by participants is a strategic and commensurate response to the repressive effects of a particular form of power. In other words, the nature of resistance is revealing of the kind of power being contested. Challenging hegemonic ideas in order to disrupt their functionality in the exercise of power reveals the workings of 'governmentality' but it also demonstrates the narrow space in which politics can be conducted safely. As Mona implies, 'active citizenship' has come to represent conformity to the state's hegemonic and passive notion of political participation.

Conclusion

This chapter has examined how young British Muslim activists negotiate the inordinate demands and contingencies placed on their citizenship in the war on terror. The dominance of 'active citizenship' in public policy has come to represent coercive demands on young Muslims to avow nationalism and demonstrate their loyalty by supporting counterterrorism policies. The chapter revealed how young British Muslim activists demonstrated differentiated intensities of resistance to such demands through 'explicit' and 'implicit' modes of cultural contestation.

Those who demonstrated more 'explicit resistance' struggled to identify with a British national identity due to feelings of exclusion and victimisation, making belonging to Britain precarious and conditional. Participants whose approach was interpreted as 'implicit resistance' described citizenship in terms that appeared to be complicit in state hegemony. It is argued that these seemingly divergent approaches represent political resistance through cultural contestation that has become one of the key technologies of power according to Foucault's (1994a, 1994b, 1995) theory of contemporary governance. Such cultural politics represent the presence of power that operates through knowledge and discourse as argued by Foucault (1994a, 1994b, 1995). This configuration of power is difficult to challenge because it is diffused in the everyday practices and taken-for-granted ways of understanding the social world. The effects of governmentality can be observed in the different ways in which young Muslim activists contest the rationalities of the state but also how they sometimes appear to reproduce them in their own discourses.

Foucault's notion of 'governmentality' (1994a) reveals how the war on terror depends upon framing security as a problem within Muslim cultures, beliefs and practices. Thus, failure to demonstrate 'active citizenship', by avowing nationalism and supporting counterterrorism, is risky for Muslims. Some participants explicitly resisted such restrictive terms of inclusion for citizenship by rejecting their national identity. Others adopted a more 'implicit' approach by subverting the given meaning of citizenship and redefining it through universal and humanistic rather than nationalistic ideals of civic duty.

Notes

1 Names of interviewees have been changed to protect their identities.

References

Ameli, S. (2002). *Globalization, Americanization and British Muslim identity*. London: ICAS Press.

Ansari, H. (2009). *The infidel within*. London: Hurst and Company.

Back, L., Keith, M., Khan, A., Shukra, K. and Solomos, J. (2009). Islam and the new political landscape. Faith communities, political participation and social change. *Theory, Culture & Society*, 26(4): 1–23.

BBC News. (2011). 'State multiculturalism has failed, says David Cameron'. BBC News online. Available from www.bbc.co.uk/news/uk-politics-12371994 [accessed 6 July 2017].

Beck, U. (1997). *The reinvention of politics*. Cambridge: Polity Press.

Briggs, R. (2010). Hearts and minds and votes: The role of democratic participation in countering terrorism. *Democratization*, 17(2): 272–285.

Brown, K. (2010). Contesting the securitization of British Muslims. *Interventions*, 12(2): 171–182.

Brown, K. and Saeed, T. (2014). Radicalization and counter-radicalization at British universities: Muslim encounters and alternatives. *Ethnic and Racial Studies*, 6(2): 1–17.

Cesari, J. (2009). Securitization of Islam in Europe. In: J. Cesari (ed.), *Muslims in the West after 9/11: religion, law and politics*. Kindle Edition. Abingdon: Routledge, 9–27.

Choudhury, T. (2007). [Forthcoming]. *The role of Muslim identity politics in radicalisation* London: Department for Communities and Local Government.

Department for Communities and Local Government (DCLG). (2010). *Literature review of attitudes towards violent extremism amongst Muslim communities in the UK*. London: DCLG.

Foucault, M. (1978). *History of sexuality Volume I*. New York: Pantheon Books.

Foucault, M. (1981). The order of discourse. In: R. Young (ed.), *Untying the text: A post-structuralist reader*. London: Routledge and Kegan Paul, 48–78.

Foucault, M. (1988). Technologies of the self. In: L. Martin, H. Gutman and P. Hutton (eds), *Technologies of the self: A seminar with Michel Foucault*. London: Tavistock Publications, 16–49.

Foucault, M. (1994a). Governmentality. In: J. Faubion (ed.), *Power. Essential works of Foucault 1954–1984 Volume 3*. London: Penguin, 201–222.

Foucault, M. (1994b). The subject and power. In: J. Faubion (ed.), *Power. Essential works of Foucault 1954–1984 Volume 3*. London: Penguin, 326–348.

Foucault, M. (1994c). Truth and juridical forms. In: J. Faubion (ed.), *Power. Essential works of Foucault 1954–1984 Volume 3*. London: Penguin, 1–89.

Foucault, M. (1995). *Discipline and punish*. New York: Vintage.

Frost, D. (2008). Islamophobia: examining causal links between the state and 'race hate' from 'below'. *International Journal of Sociology and Social Policy*, 28(11/12): 546–563.

Hall, S. (2010). Foucault: power, knowledge and discourse. In: M. Wetherell, S. Taylor and S. Yates (eds), *Discourse theory and practice: A reader*. Los Angeles; London; New Delhi; Singapore; Washington DC: Sage, 72–81.

Hindess, B. (2012). Foucaultian analysis of power, government, politics. In: E. Amenta, K. Nash and A. Scott (eds), *The Wiley-Blackwell companion to political sociology*. Chichester: Wiley & Sons, 36–46.

Home Office (2005). *'Preventing Extremism Together' Working Groups*. London: The Home Office.

Hussain, Y. and Bagguley, P. (2012). Securitized citizens: Islamophobia, racism and the 7/7 London bombings. *The Sociological Review*, 60: 715–734.

Kundnani, A. (2009). *Spooked! How not to prevent extremism*. London: Institute of Race Relations.

Kundnani, A. (2012). Radicalisation: the journey of a concept. *Race & Class*, 54(2): 3–25.

Mahmood, S. (2005). *Politics of piety: The Islamic Revival and the feminist subject*. Princeton: Princeton University Press.

Mamdani, M. (2002). Good Muslim, bad Muslim: A political perspective on culture and terrorism. *American Anthropologist*, 104(3): 766–775.

Marinetto, M. (2003). Who wants to be an active citizen? The politics and practice of community involvement. *Sociology*, 37(1): 103–120.

Mavelli, L. (2013). Between normalisation and exception: The securitisation of Islam and the construction of the secular subject. *Millennium: Journal of International Studies*, 41(2): 159 –181.

McGhee, D. (2008). *The end of multiculturalism? Terrorism, integration and human rights*. Maidenhead: Open University Press.

McGhee, D. (2010). *Security, citizenship and human rights: shared values in uncertain times*. Basingstoke: Palgrave Macmillan.

Meer, N. (2010). *Citizenship, identity and the politics of multiculturalism*. Basingstoke: Palgrave.

Modood, T. (2007). *Multiculturalism: A civic idea*. Cambridge: Polity.

Mythen, G., Walklate, S. and Khan, F. (2009). 'I'm a Muslim, but I'm not a terrorist': victimization, risky identities and the performance of safety. *British Journal of Criminology*, 49: 736–754.

Nash, K. (2010). *Contemporary political sociology: globalization, politics and power*. Second Edition. Oxford: Blackwell.

O'Loughlin, B. and Gillespie, M. (2012). Dissenting citizenship? Young people and political participation in the media-security nexus. *Parliamentary Affairs*, 65: 115–137.

Parekh, B. (2008). *A new politics of identity: political principles for an independent world.* Basingstoke: Palgrave Macmillan.

Sassen, S. (2002). Towards post-national and denationalized citizenship. In: E. Isin, F. and B. Turner (eds), *Handbook of citizenship studies*. New York: Sage, 277–292.

Soysal, Y. (2012a). Citizenship, immigration, and the European social project: rights and obligations of individuality. *The British Journal of Sociology*, 63(1): 1–21.

Soysal, Y. (2012b). Post-national citizenship: rights and obligations of individuality. In: E. Amenta, K. Nash and A. Scott (eds), *The Wiley-Blackwell companion to political sociology*. Chichester: Wiley & Sons, 383–393.

Spalek, B. (2010). Community policing, trust, and Muslim communities in relation to 'new terrorism'. *Politics and Policy*, 38(4): 789–815.

Thomas P. (2016). Youth, terrorism and education: Britain's PREVENT programme. *International Journal of Lifelong Education*, 35(2): 171–187.

Vertigans, S. (2010). British Muslims and the UK government's 'war on terror' within: evidence of a clash of civilizations or emergent de-civilizing processes? *The British Journal of Sociology*, 61(1): 26–44.

Vertovec, S. and Wessendorf, S. (2010). Introduction: assessing the backlash against multiculturalism in Europe. In: S. Vertovec and S. Wessendorf (eds), *The multiculturalism backlash: European discourses, policies and practices*. Kindle Edition. London: Routledge, 1–31.

Wiktorowicz, Q. (2005). *Radical Islam rising: Muslim extremism in the West*. Lanham, MD: Rowman & Littlefield.

Part IV

Resisting and creating new public spheres

Chapter 10

What future for young people's artistic activism?

Jane McDonnell

Introduction

This chapter centres on one particular aspect of youth activism and its governance, i.e. the role of arts-based strategies and the aesthetic within young people's political protest. In doing so, it addresses one of the key themes of the book: the paradox that liberal-democratic states ostensibly committed to democratic values now repress various forms of political action, including the use of artistic strategies within young people's protest. The chapter first outlines both the increased use of artistic strategies within youth activism, and the increased surveillance of such activity in established democracies. It goes on to argue that this repression occurs not only through overt measures of criminalisation and control but also, in a Foucauldian sense, through the more diffuse workings of power within neo-liberal states (Foucault 1977; Lemke 2002). This highlights a further paradox: that the systems and institutions through which such power operates are also the very sites in which political dissent can have most effect.

Drawing on a Rancièrian (2004, 2007) reading of the relationship between politics and aesthetics – in which the very act of politics can be read as an aesthetic activity in itself – the chapter goes on to analyse one recent instance of political protest amongst young people in the UK. While the protest itself was not explicitly artistic, the analysis illustrates how young people's political dissent can disrupt existing power structures through an aesthetic reconfiguration of roles in which the instruments and operations of surveillance are turned on themselves. Finally, I argue for a shift away from 'artistic activism' towards a greater recognition of the aesthetics of political dissent. To the extent that artistic or 'aesthetic activism' is understood as an ally in making the systems of power apparent, it is a vital part of youth politics. This also requires a re-visioning of aesthetic activism based on a view of the relationship between art and politics, in which the two are seen as inextricably and *unpredictably* bound up with each other. The principal argument advanced in this chapter, then, is that the way in which we understand the power of art and the aesthetic within young people's activism can be broadened to incorporate activity that aesthetically challenges the boundaries of the political – and therefore also represents a challenge to any given political 'status quo'. Furthermore, when viewed in this way, aesthetic activism can be understood as one of the most important ways in which young people can continue to resist and express dissent within and against an environment of increased surveillance.

Young people, political protest and the governance of artistic activism

Young people's involvement in political protest in established democracies has seen a dramatic rise since the financial crisis of 2008, with high numbers of young people involved in civil unrest and grass-roots activism (Sloam 2014; Loader *et al.* 2014). In the UK, prominent examples include the Occupy movement, mass protests over rises in student fees (Rheingans and Hollands 2013) and the large numbers of young people involved in the campaign to elect Jeremy Corbyn as leader of the Labour Party (Diamond 2016). Despite the persistence of arguments about young people's perceived apathy based on a low voter turnout, many have highlighted the innovative ways in which young people *are* engaged in politics. Sloam (2014), for example, has argued that youth activism has become an important feature of the political landscape in Europe, quickened by access to social media.

Alongside the use of social media, an important dimension of contemporary youth politics has been the use of artistic imagery, methods and approaches within political protest. Examples include the adoption of imagery from popular culture, including the adoption of Guy Fawkes masks alluding to Alan Moore's comic, then the film, *V for Vendetta*, first by 'Anonymous' and then by other movements such as Occupy. Elsewhere, the actions of Pussy Riot, making use of music and performance art within political protest in Russia, have drawn attention. In the US, Barnard (2011) has analysed how performative strategies such as street theatre within the 'freegan' movement have been used to create images that challenge the dominant discourses of capitalism. As well as the use of artistic methods and images within protest, there has also been marked interest in the aesthetic dimensions of new forms of activism. Hatuka (2016), for example, has explored how movements such as Occupy, are reconfiguring urban spaces. Hatuka's (2016) work is particularly helpful in highlighting the diffuse ways in which aesthetics play an important role in politics.

However, at the same time that there has been an upsurge in young people's activism – artistic and otherwise – there has been an increase in measures used by liberal-democratic states to contain, regulate and repress such activity. In the UK, recent legislation to control political dissent has been introduced under the counterterrorism CONTEST and PREVENT agendas. In addition, Power (2012) has highlighted how new forms of police control, such as 'kettling' at mass demonstrations, have resulted in the criminalisation of protest and protesters. Bessant (2016) has taken the example of digital denial of service activities to illustrate a broader trend of increased criminalisation of new forms of youth political activism. Particularly important is her illustration of *how* such forms of activism come to be criminalised, e.g. through trivialisation and downplaying the political import of protest activities. Similar trends can be detected in the regulation and control of artistic forms of activism. Young (2010), for example, has documented attempts to control, manage and criminalise street art as a political force, particularly graffiti.

Such overt mechanisms of repression are a significant cause for concern. However, it is also important to recognise that governmental power works in diffuse ways, often within the systems and structures that are intended to articulate young people's political agency. Drawing on Foucault's concept of governmentality (Foucault 1977; Lemke 2002), the surveillance and governance of young people's political activism can be read as a series of technologies and operations through which neo-liberal states govern. Ball (2009, 2012) has charted how such operations feature in education, through neo-liberal systems of accountability, performativity, marketisation and privatisation. These serve to repress dissent through self-governance. The reductive application of 'student voice' in schools (Rudduck and Fielding 2006), for example, can be read as a constraining technology that offers the appearance but not the experience of real power. The problem is not only that young people's political dissent is being overtly criminalised, but also that it is being diverted and suppressed within a configuration of power that operates through the very systems and institutions designed to give them voice – schools, student councils, the broader education system and its accountability to the public via the press.

Nevertheless, such configurations are always open to disruption. As Butler (1993) has argued, discursive configurations of power can be both constraining *and* enabling, and the various 'slippages' that occur in recitations of discourse can disrupt existing power structures. She has written on the politics of performative subjectivity, as a process of '*dis*identification', which is 'crucial to the re-articulation of democratic contestation' (1993: 4). Her work has also been taken up widely within the sociology of education in relation to schools and other educational contexts (e.g. Youdell 2006; Hey 2006). Ball (2016), following Butler's theoretical work (1990, 2005) has recently recognised the importance of subjectivity as a site of political struggle. In the discussion that follows, I aim to illustrate how Rancière's (1999, 2004, 2006, 2007) work helps to think through the role of art and aesthetics within such struggle, and to argue for a shift away from the concept of artistic activism towards a notion of 'aesthetic activism'. Furthermore, I argue that such activism is better placed to address the forms of surveillance and repression that operate through the neo-liberal operations of governmentality. Indeed, the aesthetic work of making such things visible may be one of the few ways in which young people can continue to resist and take action that leads to real political change.

Theorising artistic activism

The kinds of activities outlined above, e.g. the adoption of cinematic tropes and the use of street performance, can be read as part of what Chantal Mouffe (2007) has described as 'artistic activism'. Based on a specific understanding of democratic politics as an 'agonistic' sphere of contest in which radically different visions of community can come into positive confrontation with each other, Mouffe (2007) has argued that art can play an important role within political struggle. Mouffe views public spaces as 'the battleground where different hegemonic projects are confronted, without any possibility of final reconciliation' (2007: 3) and sees artistic strategies as an important weapon in the democratic arsenal because they disrupt the smooth presentation of politics in terms of consensus. Mouffe's work is particularly significant in painting a picture of how art and politics can be seen as intimately bound up with each other in artistic forms of political activism. Rather than conceptualising art and politics, 'in terms of two separately constituted fields, art on one side and politics on the other, between which a relation would need to be established' (Mouffe 2007: 4), Mouffe sees the two as being intimately connected at the level of a symbolic order. For Mouffe, art has a political dimension, and politics an artistic dimension, such that artistic practices can contribute to disrupting the dominant hegemony of a smooth consensus in the public sphere. She argues in particular for the capacity of 'critical art' in this respect as it contributes to the construction of new subjectivities.

The view of artistic activism that Mouffe outlines has been significant in moving towards an understanding of the role of art within political activism not as an adjunct or gimmick but as a very real and important part of political struggle. Working with a rather Gramscian concept of a symbolic or cultural hegemony that can only be challenged in symbolic terms, Mouffe outlines the way in which such symbolic artistic activism can be taken up within radical democratic politics. Her work has also been influential in the field (see, e.g. Bell and Desai 2011; Stavrakakis 2012), although the work of arts theorists on the social, political and educational role of art (see, e.g. Kester 2011; Bishop 2012) have also been significant. Mouffe's approach is also not without its problems. Particularly in her argument about allying artistic activism to other, more traditional forms of political action and protest, Mouffe's characterisation in a sense downplays the aesthetic significance of artistic forms of political protest. Equally, this element of Mouffe's argument relies on a teleological and somewhat monolithic view of politics as a 'war' to be won by 'parties and trade unions'. Crucially, Mouffe argues that artistic activities cannot alone achieve a subversion of the dominant hegemony, and sees the need for artistic activism to be combined with more traditional political forms.

Rancière's views on the relationship between art and politics are also dependent on a very specific understanding of politics and democracy. However, Rancière sees democracy as a disruptive and dynamic movement that is embodied in 'politics' – i.e. real or true politics – which he opposes to the 'para-politics' of representative democracy or the 'metapolitics' (1999) of ideological movements such as Marxism. The dynamism and fluidity of such democratic politics is expressed in Rancière's claim that democracy 'is only ever entrusted to the constancy of its specific acts' (2006: 7), a characterisation which sits within a broader historical account of democracy. For Rancière, there exists within all political settlements, since the original emergence of democracy in Athens, a constant struggle between a 'police' logic and 'politics'. The 'police' or 'police logic' is that which partitions and maintains the distribution of roles within the polity according to 'social competences', and 'politics' is that which disrupts this partition, insisting instead on the more chaotic (and democratic) rule of anyone and everyone (Rancière 2006: 55). For Rancière, democracy is a process of political subjectification that involves specific disputes over the partition of roles in society, made in the name of equality and exposing the underlying equality that exists in all *political* systems. Because this entails the disruption and redistribution of roles between the private and public spheres, democracy also involves the 'enlarging of the public sphere' (2006: 55) by bringing private roles (such as that of women), relations and disputes (such as those between employers and employees) into the public realm. These contradictions and tensions within any given social and political order become, within Rancière's view, their Achilles heel. Any political order, benign or malevolent, is only ever one specific democratic dispute away from a dramatic redistribution. Indeed, such disputes and actions have an inherently dramatic, aesthetic dimension; they stage the tensions and contradictions within the police order in inventive ways.

The relationship between art and politics in Rancière's work centres on his concept of 'the distribution of the sensible' (2004: 12), which he defines as 'the system of a priori forms determining what presents itself to sense experience' (2004: 13). This is the partition of spaces, places and activities within a community, which renders some things visible, audible and thinkable while others are not. It constitutes an aesthetic field of possibilities from amongst which any political distribution of the community finds its form. Rancière expresses this most clearly in his claim that '[p]olitics revolves around what is seen and what can be said about it, around who has the ability to see and the talent to speak, around the properties of spaces and the possibilities of time' (2004: 13). Politics can literally render new possibilities visible, and art can create new ways of being that disrupt a given distribution of places within the political community. He argues that '[t]he channels for political subjectivisation are not those of imaginary identification but those of "literary" disincorporation' (Rancière 2004: 40) and claims that '[m]an is a political animal because he is a literary animal who lets himself be diverted from his "natural purpose" by the power of words' (Rancière 2004: 39). For Rancière, artistic practices are related to politics because, through a particular form of 'aesthetic' equality (2004: 55), they create channels for subjectification that can disrupt and reconfigure the distribution of roles, places and occupations within a community. When they do this in a way that disrupts and displaces a distribution based on a 'natural' logic of inequality, they share a logic and common purpose with democratic politics.

The contributions of both Mouffe and Rancière in this area offer ways of thinking through the relationship between art and politics in which the two are seen not as separate spheres, but rather as constitutive of each other. They both also suggest that the relationship between art and politics exists at the level of subjectivity, and that art can be a factor in the disruption of political realities by providing new subjectivities and new ways of being. However, their arguments also diverge in important ways. Whereas Rancière maintains that the relationship between particular art works and their democratic effects is always unstable, Mouffe argues that artistic activities can and should be allied to particular political projects. Equally, there is an unpredictability to the relationship between art and democratic politics in Rancière's work, which is not present in Mouffe's argument. For Rancière, democracy can occur 'whenever a community with the capacity to argue and make metaphors is likely, at any time and through anyone's intervention, to crop up' (Rancière 1999: 60). For Mouffe, there has to be a strategy for making artistic interventions. For Rancière, the contribution of art to political subjectification is something that cannot be predicted, much less planned from a within a political programme. Like democracy itself, such contributions can be created and taken up by anyone at any time, without knowing in advance where they might lead.

While Rancière does not use the term 'artistic activism', his discussions of the relationship between art and politics can be read as an important way of adding to this concept and redirects attention from the use of art as a tool within political activism towards an awareness of the aesthetics of political activism and protest. In the section that follows, I outline one such instance of political protest amongst young people in the UK, and analyse this via Rancière's theory of the relationship between politics and aesthetics in order to illustrate how 'aesthetic activism' can work to disrupt and displace the apparatus of surveillance through which young people's political activity is controlled and stifled.

A Rancièrian reading of a political action amongst young people in the UK

The instance of direct political action I wish to explore here involved the boycott of their school canteen by a group of secondary school pupils at a school in the UK, organised in response to steep price rises, which followed the redevelopment of the school as part of the Public Funded Initiative scheme. This scheme was designed to facilitate funding to public services and infrastructure through partnership with private enterprise. It was first introduced in 1992 by the Conservative government of 1979–1997 but was then enthusiastically embraced by New Labour (Flinders 2005). Ball (2007) has described it as one of the most radical forms of privatisation of public services that 'reworks the landscape of the public sector' (2007: 47). The description below is illustrated with reference to data from a research project funded by the Education and Social Research Council, carried out over a period of two years with young people who had recently taken part in a gallery education project and were attending the same secondary school. The project centred on the impact, over time, of their participation in the project set against their broader experiences of democracy, education and the arts. In the account that follows, the names of all the participants have been changed to preserve anonymity.

One of the participants, Emma, described how the boycott had been undertaken in response to steep price rises imposed by the private catering company now running the canteen. Subsequent interviews with both Emma and other participants allowed for a more detailed exploration of the boycott. It became clear that the young people had decided on this course of action after first raising their concerns with the catering company via the student council and the perceived unwillingness of the company to listen to their concerns and take the students seriously, as the excerpt from an interview with Emma, below, illustrates:

> [The staff] said, 'There's nothing we can do about it' and I think they were quite rude to them [the student reps] as well, so … and then they said, 'Well that's what you end up … we end up going on strike'.
>
> (Emma)

The fact that the company did not appear to take the students seriously was an important factor in precipitating the action. From the students' perspective, not only were the company unprepared to act on their concerns, as voiced through the appropriate democratic channel of the student council (which the students characterised as normally concerning itself with trivial and unimportant matters), but they also did not appear to treat the students with due respect. When discussing the boycott, it became clear that the students were also motivated by a sense of injustice at being denied what they saw as their entitlement to controlled prices because of a school rule about not being allowed outside the premises at lunchtime:

> From the beginning when [the company] came into the school when the new school was built, the prices had gone up by like, oh I can't remember what it is now but it was a big amount, especially for us who are like … and our prices were meant to be … because obviously we're tax free and as well we're meant to be a lot lower because it's a school sort of thing and like … but then we had to pay the prices and everyone was like, 'If we were allowed to go out of school to like Tesco's for our lunch it would be so much cheaper'.
>
> (Emma)

What is most telling here is that the decision to take the action was not only, or even primarily, because of the price rises. Rather, it was a result of the students' sense of injustice at being treated simultaneously as both consumers and children, but paradoxically enjoying the benefits of neither status. In addition, they felt that the staff of the catering company had spoken down to them. In other words, the boycott was not primarily an economic dispute over money but rather a political dispute over the students' status as political beings caught up in a larger conflict between public and private interests.

In Rancièrian terms, the boycott was really about the staging of a very real dispute over the boundaries between the public and the private sphere, and the distribution of roles, spaces and positions therein. Furthermore, this political action had an important aesthetic dimension. The boycott was staged as a one-day protest and involved the majority of the student body. It caused a considerable amount of disruption and attracted attention in the local press, with journalists coming to report on the students' action. In making sense of the boycott and its significance, some of the participants reflected on the aesthetic dimensions of this protest action. Claire, for example, talked about how seeing journalists barred at the gates of the school drew her attention to the fact that the school and its grounds were now owned by a private company:

> Erm, but no, it was just so funny because the newspapers came and [the company], who are the people who own it, they wouldn't let them in because they own the school and so there were all these people outside who had been barred out and then everyone was sort of just like stood on the hill. I don't know, it was just the most surreal thing, it was really funny.
>
> (Claire)

For Claire, the boycott led to a new awareness of the political circumstances surrounding the school's ownership and the public-private partnership behind it. Interestingly, coming to this awareness was also an aesthetic experience. The visual impact of the boycott – the 'surreal' and 'funny' sight of the local newspaper reporters being barred at the gates of the school – led to this new awareness. The sight of nearly all the students gathered together also made an impact on Claire, and caused her to see the school community, and her place within it, differently:

> It was just so strange because it was literally like three quarters of the school just all in the courtyard and there were so many people. I'd never seen like the whole school together as well, so it was quite nice how everyone did join in and support it.
>
> (Claire)

This shift in perception about the political realities affecting the students' lives is illustrative of what Rancière means in arguing that democratic politics, and political action, causes a rupture in the 'distribution of the sensible' (2004: 12). Such actions always involve an aesthetic dimension in what they make visible, doable and possible. In Claire's case, the boycott literally caused her to see things in a new way, visually highlighting dormant realities about the ownership and governance of the school, as well as the impact of the students' protest and their existence as a collective body, capable of real political action. Claire's reflections demonstrated the aesthetic dimensions of the political act of taking part in a boycott. For Emma, taking part in the planning and implementation of the boycott caused her to reflect on the use of artistic and aesthetic strategies within political protest. Considering the effectiveness and impact of the boycott, Emma reflected on the kinds of artistic strategies that might otherwise have been employed:

> In my opinion, it would've been so much more effective if we'd all just like stood or like sat or even like gone into the canteen … it would have been more effective if everyone had brought their packed lunch and everyone had gone into the canteen and sat there in silence it would have had the most effect.
>
> (Emma)

This consideration of potential aesthetic and artistic strategies also appeared to be related to Emma's changing attitudes to art at the time, having recently been involved in a gallery education project in which she had encountered new kinds of artistic practices and begun to think differently about what counts as art:

> I don't know it's like, it's like different. I suppose there's always been like a sort of like stereotypical sort of form of art and this like, which is like sort of paintings and that kind of thing, it's just like completely different and you have to like, you look at it and you think like, or well a lot of people would think, 'That isn't art', then it's like, it's nice to sort of, especially if you read about it and like find out like what the artist was thinking and why like … because when you, if you looked at it straight away, like you don't always think, like you wouldn't always think that's a piece of art work but then you sort of read into it and you think, 'Oh yeah, I see why that thing's, what that's representing' and like, stuff like that.
>
> (Emma)

Emma clearly had an appetite for engaging in artistic activism, in the sense of a deliberate alliance of artistic strategies to discrete political ends, as envisioned by Mouffe (2007).

The aesthetics of political protest: making visible the contradictions of repression

The political action described above is one small example of protest, and did not involve the deliberate use of artistic strategies. However, a Rancièrian reading of this incident helps to highlight some of its aesthetic dimensions and indicates some important implications for the way in which we think of artistic and perhaps 'aesthetic' activism. First, it illustrates the unpredictable nature of the relationship between art and politics. It was not primarily in the deliberate use of artistic strategies that the boycott was a significant act both politically and aesthetically, although Emma's musings on the potential efficacy of artistic and aesthetic strategies are also important. Rather it was in the aesthetic dimensions of the political action itself – in the 'surreal' and strange apportioning of space and bodies in and around the school – that it had a dramatic impact, as Claire's reflections on the event clearly demonstrate. Indeed, the term 'aesthetic activism' might be more helpful in understanding the nature and impact of such political protest. This is not to suggest that the deliberate use of artistic strategies within political protest is not also important. Furthermore, the young people's reflections point to the possibility that arts participation and engagement with the arts can be an important site in which an awareness of the aesthetics of political protest are tested and heightened.

Second, this Rancièrian reading of a political action demonstrates how the suppression of political activism is something that occurs not only through legislation and policing but also through the systemic arrangements and institutions that can render people's political agency and concerns invisible. It was through the strange, internal tension and contradiction between definitions of the students as simultaneously both consumers, and children under the care of the state, that the company was able to impose unreasonably high prices in the canteen. This was possible because such a definition of the students rendered them invisible as political subjects; they were consumers not citizens, children not full adult members of the polity.

Compounding this, each definition neutralised the other: as children under the care of the state, they were not free to exercise their spending power elsewhere; as consumers, they were stripped of their rights to subsidised or free meals. This strange double bind was, of course, a result of the deeper, internal tensions and contradiction inherent in the delivery of state education via a public-private partnership. Happily, however (if we are concerned with young people's ability to take real political action resulting in real political change), it is that very contradiction and double bind that allowed the students to take effective action and stake a legitimate claim to political visibility – first via the student council, then via direct political action in the form of a boycott. It is precisely these contradictions within the institutions and systems that suppress political action, that Rancière's (2006) work shows us are the real terrain and battleground for genuine politics, understood as an aesthetic matter (2004, 2007).

Finally, and perhaps most importantly, the discussion illustrates how such instances of political action can turn the instruments of surveillance on themselves.

Returning to the concept of governmentality (Foucault 1977; Lemke 2002), its operation in and through the education system (Ball 2009, 2012) outlined in the introduction, the young people can be seen as caught up within a whole set of instruments, technologies and apparatus of surveillance and governance of their political activity. Interestingly, though – and returning here to the discussion of how such configurations can be both constraining and enabling (Butler 1993) – the young people were able to make creative use of some of the apparatus of governance. The student council is an interesting example in this respect, since such fora are intended to give political power to young people within schools but often end up reproducing existing constraints (Rudduck and Fielding 2006). The young people were able to make creative use of the student council to stage the boycott, despite it being viewed as a forum that typically concerned itself with unimportant matters. The way in which the young people's action disrupted the roles and relationships of governance was also very visually illustrated in the displacement of pupils, press and school staff within and outside the school grounds during the boycott itself. The company employees became the objects rather than the subjects of surveillance in relation to the students, even while they remained the surveyors of the press. The journalists too, and their bodily presence on the school site, became the object of surveillance not only for the school's owners but also for the students. Claire's characterisation of this scene as 'funny' and 'surreal' aptly captures both the absurdity and the drama of the whole series of shifts, exchanges and displacements of roles that the boycott affected.

In this chapter, I have outlined the increase in young people's participation in new forms of political protest, including those that deploy artistic and arts-based strategies. I have further illustrated the way in which such forms of youth politics have been the subject of increased surveillance in liberal-democratic states. Taking a Foucauldian view of surveillance and governmentality, in which power operates through a series of technologies, including within the education system (Ball 2009, 2012), I have highlighted the more diffuse and oblique ways in which youth politics is governed and controlled. Based on a reading of Butler (1990, 1993, 1997) and Rancière (1999, 2004, 2006, 2007), I have further argued that such technologies can be disrupted and reconfigured through political action that is also aesthetic. Finally, I have applied this argument to a specific instance of political protest amongst young people in the UK to illustrate precisely *how* this can happen through an aesthetic disruption and reconfiguration that turns the instruments of governance and surveillance on themselves. The constant surveillance of school performativity through the press, for example, can be inverted; those who are normally the subjects of surveillance can become its objects. Furthermore, the operations through which young people's political subjectivity are often neutralised can become reanimated when taken seriously, in an unpredictable fashion, through anyone's intervention. When we understand artistic activism as encompassing the staging of political acts that expose the contradictions and tensions in the apparatus of suppression and surveillance – as an 'aesthetic activism' perhaps – it is possible to see such activism as a vital and still powerful weapon within the arsenal of youth politics. What is particularly powerful about such activism is that it targets not only the discrete political issues of concern to young people, but also the very systems through which political protest and youth politics are governed. Such acts are particularly helpful in subverting surveillance and shifting perception as a way of bringing about real political change. This contribution complements those analyses of youth politics that highlight the repression of political dissent through overt forms of surveillance. It both recognises the way in which such repression impacts specifically on artistic activism and offers an alternative (and hopeful) vision of an 'aesthetic activism' – one that can challenge more diffuse forms of repression that operate in an age of surveillance.

References

Ball, S. (2007). *Education plc: understanding private sector participation in public sector education*. London: Routledge.

Ball, S. (2009). Privatising education, privatising education policy, privatising educational research: network governance and the 'competition state'. *Journal of education policy*, 24(1): 83–99.

Ball, S. (2012). *Foucault, power, and education*. London: Routledge.

Ball, S. (2016). Subjectivity as a site of struggle: refusing neoliberalism? *British Journal of Sociology of Education*, 37(8): 1129–1146.

Barnard, A. (2011). 'Waving the banana' at capitalism: political theater and social movement strategy among New York's 'freegan' dumpster divers. *Ethnography*, 12(4): 419–444.

Bell, L. and Desai, D. (2011). Imagining otherwise: Connecting the arts and social justice to envision and act for change: special issue introduction. *Equity & Excellence in Education*, 44(3): 287–295.

Bessant, J. (2016). Democracy denied, youth participation and criminalizing digital dissent. *Journal of Youth Studies*, 19(7): 921–937.

Bishop, C. (2012). *Artificial hells: Participatory art and the politics of spectatorship*. London: Verso.

Butler, J. (1990). *Gender trouble. Feminism and the subversion of identity*. New York: Routledge.

Butler, J. (1993). *Bodies that matter: on the limits of 'sex'*. London: Routledge.

Butler, J. (1997). *Excitable speech: A politics of the performative*. New York: Routledge.

Butler, J. (2005). *Giving an account of oneself*. New York: Fordham University Press.

Diamond, P. (2016). Addressing the performance of UK opposition leaders: Jeremy Corbyn's 'Straight Talking, Honest Politics'. *Politics and Governance*, 4(2): 15–24.

Flinders, M. (2005). The politics of public-private partnerships. *The British Journal of Politics and International Relations*, 7(2): 215–239.

Foucault, M. (1977). *Discipline and punish. The birth of the prison*. New York: Pantheon.

Hey, V. (2006). The politics of performative resignification: translating Judith Butler's theoretical discourse and its potential for a sociology of education. *British Journal of Sociology of Education*, 27(4): 439–457.

Hatuka, T. (2016). The challenge of distance in designing civil protest: The case of Resurrection City in the Washington Mall and the Occupy Movement in Zuccotti Park. *Planning Perspectives*, 31(2): 253–282.

Kester, G. (2011). *The one and the many: contemporary collaborative art in a global context*. London: Duke University Press.

Lemke, T. (2002). Foucault, governmentality, and critique. *Rethinking Marxism*, 14(3): 49–64.

Loader, B., Vromen, A. and Xenos, M. (2014). The networked young citizen: social media, political participation and civic engagement. *Information, Communication & Society*, 17(2): 143–150.

Mouffe, C. (2007). Artistic activism and agonistic spaces. *Art & Research*, 1(2): 1–5.

Rancière, J. (1999). *Disagreement: politics and philosophy*. Minnesota, MN: University of Minnesota Press.

Rancière, J. (2004). *The politics of aesthetics: The distribution of the sensible*. London: Continuum.

Rancière, J. (2006). *Hatred of democracy*. London: Verso.

Rancière, J. (2007). *The future of the image*. London: Verso.

Rheingans, R. and Hollands, R. (2013). 'There is no alternative?': Challenging dominant understandings of youth politics in late modernity through a case study of the 2010 UK student occupation movement. *Journal of Youth Studies*, 16(4): 546–564.

Rudduck, J. and Fielding, M. (2006). Student voice and the perils of popularity. *Educational Review*, 58(2): 219–231.

Sloam, J. (2014). 'The outraged young': young Europeans, civic engagement and the new media in a time of crisis. *Information, Communication & Society*, 17(2): 217–231.

Stavrakakis, Y. (2012). Challenges of re-politicisation: Mouffe's agonism and artistic practices. *Third Text*, 26(5): 551–565.

Power, N. (2012). Dangerous subjects: UK students and the criminalization of protest. *South Atlantic Quarterly*, 111(2): 412–420.

Youdell, D. (2006). Subjectivation and performative politics—Butler thinking Althusser and Foucault: intelligibility, agency and the raced–nationed–religioned subjects of education. *British Journal of Sociology of Education*, 27(4): 511–528.

Young, A. (2010). Negotiated consent or zero tolerance? Responding to graffiti and street art in Melbourne. *City*, 14(1–2): 99–114.

Chapter 11

Effects of the regime in Malaysia on youth political participation

Norhafiza Mohd Hed

Introduction

A significant body of research testifies that political dissent, especially amongst young people, is regularly subject to surveillance and criminalisation (Bessant 2016; Fernandez 2008). There is some surprise amongst observers when those liberal democracies that guarantee civil liberties such as freedom of expression, freedom of assembly and human rights, repress legitimate dissent. And yet, the use of pre-emptive control tactics against young protesters, including 'antiterrorism' laws, crowd control policing and non-lethal weapons, is apparent in many established democracies. In 2010, for example, the student demonstration 'Fund our Future' in the UK and the G20 protest in Toronto saw clashes between protesters and police in riot gear, leading to kettling and mass arrests. Admittedly, such cases are minimal compared to those in authoritarian states. Authoritarian regimes have long criminalised dissent to stay in power. But even here, prior to youth uprisings pushing for democratic change, such as the campaign to topple Suharto in Indonesia in 1998 and the Arab Spring in 2011 (Lucan Way 2014), increasing numbers of young people have been targeted and repressed by such regimes. How do young people living under such regimes respond to oppressive rule? Does a culture of political repression constrain young people's engagement in politics?

Little research has been conducted into the impact of authoritarian regimes on youth political participation, particularly in Malaysia. Indeed, Malaysia raises questions about the political role of young people confronted with the possibility of pursuing democratisation while facing a government committed to repressing such a project. This was the case in the late 1990s when there were significant attempts at democratic reform, especially during the *Reformasi* period of 1998, which saw the regime deploy several repressive mechanisms. In this chapter, I ask what effect did the efforts of the Malaysian government to repress democratic dissent have on the disposition of young Malaysians to engage politically. Drawing on 20 interviews with young Malaysians, this chapter explores the effects of Malaysia's authoritarian regime on young people's politics and what, if any, effect this had on their engagement in politics.[1]

I begin by sketching how the regime in Malaysia actively sought to 'depoliticise' and criminalise young people's dissent by briefly examining cases involving young people. I then discuss the concepts of state repression and youth political participation and use both concepts as a framework to understand whether state repression and the criminalisation of democratic dissent have affected youth political participation, and how young Malaysians respond to attempts to govern their politics in repressive ways.

How Malaysia's regime regulates youth politics

As a semi-democratic regime, the criminalisation of dissent in Malaysia, particularly to regulate young people's politics, is a common phenomenon and has a long history. The implementation of a broad range of legal sanctions, coupled with the suppression of civil society groups and certain kinds of political activism amongst the young – such as protests – are part of how the Malaysian government tightens control over young people expressing political discontent, and it helps to secure the power of elites (Barraclough 1985). It is therefore crucial to provide a short background of the Malaysian government's attempts to regulate young people's political activities.

It is difficult to be exact about when the government began using repressive measures against young Malaysians, but this was clearly apparent in the 1970s during the era of 'student activism'. At the time, young people (largely students) were protesting to champion the rights of poor and marginalised communities. The well-known Baling incident of 1974 is an example that shows how the ruling regime took pre-emptive controls against students who were critical of government policies by arresting them under the Internal Security Act (ISA), which permits detention without trial (Karim and Hamid 1984). Not long after the protest, the government amended the Universities and University Colleges Act (UUCA) 1971.[2] With the enforcement of the Act, students were no longer allowed to join or support any political party, trade union, or participate in off-campus activities, and all student organisations were dissolved. These amendments initially slowed down the pace of student protest activism and sought to 'depoliticise' young people by limiting their political freedom and activities.

Youth activism appears to be on the rise primarily in the *Reformasi* of 1998, prior to the dismissal of the former deputy prime minister, Anwar Ibrahim, from the government. Anwar galvanised large numbers, particularly young people, who came onto the streets to protest his dismissal and call for the prime minister's resignation (Weiss 2006). The protests met with increasingly harsh criminalisation from the government, which included the use of mass police force and non-lethal weapons such as water cannon and tear gas against young dissenters. Despite the *Reformasi* movement waning, youth activism energetically continued to advocate for democracy and human rights, but the regime continued to detain young people. In June 2001, the 'anti-ISA' protest saw several of its young key players arrested and suspended from universities (Weiss

2011). After a decade of resisting and protesting against indefinite detention without trial, only in 2011 did the government repeal the Act and replace it with the Security Offences (Special Measures) Act 2012, which aimed at maintaining public order and national security. These tactics were repeated against students protesting to abolish the UUCA, but the government only slightly amended Section 15 of the UUCA to allow students to join political organisations, including political parties, outside campuses. This amendment did not appease students, since they still had limited autonomous power and little academic freedom (Suhaimi 2013).

More recently, one of the laws most often used by the government to silence critics is the Sedition Act 1948. The Sedition Act, under Article 3(1), criminalises any speech, actions, words and publications that contain a 'seditious tendency' to encourage hatred or excite dissatisfaction about any ruler or government, to raise discontent amongst the population and to promote hostility between races and social classes. It seems that the number of individuals who have been charged under this law, including young people, has increased significantly from around 30 cases in 2012 to 91 cases in 2015 (Amnesty International 2016). For example, a well-known youth activist, Adam Adli, was spared a one-year prison sentence but fined RM 5,000 (USD 1,210) under Section 4(1)(b) of the Sedition Act after allegedly inciting the public to overthrow the government through street protests during a political forum in 2013.

Although the government announced its commitment not to censor internet communication, the enactment of the Communications and Multimedia Act (CMA) 1998, mainly Section 233(1)(a), acts as a mechanism of surveillance and criminalisation against dissent. Section 233(1)(a) undermines freedom of speech by criminalising a person who uses network services and network facilities to transmit any communication that is deemed to be offensive with intention to annoy, threaten and abuse another person. This Act has been used frequently against youth activists. For instance, Khalid Ismeth was charged with eleven counts under Section 233 of the CMA and another three counts under Section 4(1) of the Sedition Act for posting comments on Facebook that were allegedly offensive against the Johor royalty and the Malaysian police (Amnesty International 2016). Another recent case involves the graphic artist and activist Fahmi Reza, who is known for posting on social media a caricature of Prime Minister Najib Razak as a clown. Due to this image, which was widely shared across social media, Fahmi Reza's Twitter account was placed under surveillance by the police and he was charged on two separate counts under the same section of the Act, which carries a maximum of two years in jail or RM 50,000 (USD 12,178) fine, on conviction.

Against this backdrop, we can see that criminal laws and political violence are used by the government as a mechanism for regulating youth political activism in Malaysia. The government has also expanded its use of repressive laws to criminalise not only offline but also online activism, including political satire. I now turn to focus on the different forms of state repression and criminalisation and their implication on youth political activism in Malaysia.

Criminalisation of dissent as state repression

Much literature discusses the criminalisation of political dissent in the context of state repression, mainly problematising the role of the state in governing dissent. Scholars of social movements such as Hibbs (1973), Tilly (1978) and Snyder (1976) emphasise the role of state coercion and collective action in shaping the potential of political movements to open up democratic space in a political culture. Tilly (1978) provides an explicit theoretical framework for understanding the political effects of state repression and is one of the key scholars of social movements who has developed political opportunity theory (Tarrow 1998). Theorists like Tarrow argue that the actions of activists, such as those in pro-democracy movements, are dependent on the presence or absence of specific political opportunities. Those opportunities may take the form of a decline in repression or a division within the ruling elite which decrease the likely costs of collective action. This theoretical approach seems applicable to the examination of political participation more generally.

For Goldstein (1978), state repression entails the use of physical sanction, whether actual or threatened, against an individual or organisation within state jurisdiction in order to impose a cost on the target or to prevent activities deemed to be suspect or threatening to the regime. Repression also occurs when governments raise the financial costs of mobilisation and collective action, when they disrupt access to communication and to resources (Tilly 1978), and when they limit the freedoms of individuals and groups demanding political change (Earl 2011). Consequently, the increase of dissident behaviour leads to an increase in the state's repressive efforts, since repression and dissent seem directly interrelated with one another (Davenport 1995; Tilly 1978). In other words, a 'state's repressive propensity' (Hibbs 1973) is determined by the behaviour of dissidents (violent or non-violent). On this view, dissent signifies those efforts to challenge and change the status quo, while repression represents those efforts to alleviate and stabilise dissent (Davenport 2009).

Different forms of repression are used by the state to silence dissent. First, state repression can mean restrictions on civil liberties, so-called 'soft' repression. This refers to a government's policies to restrict freedom of speech, religion, movement and work (Hibbs 1973). Restrictions on civil liberties, as argued by Davenport (2007), specifically aim to change attitudes by raising the cost of dissent. Another type of repression is known as violent or 'hard' repression, which encompasses the use of murder, surveillance, political terror, imprisonment, torture and mass killing (Davenport and Armstrong 2004). Violent repression is generally used by a government to threaten and intimidate dissenters. Both types of repression are closely related to this study, particularly when considering the repressive measures used by the regime in Malaysia to silence youth dissent and criminalise certain forms of political activism.

Repression is a strategic option for a regime wanting to govern dissent, but why do some regimes repress differently to others? Davenport (2007), amongst others, emphasises that the degree of repression and perceived threats by the state or elites might vary across different types of regime. For example, liberal-democratic regimes are less likely to exercise repression and are committed to tolerate certain forms of activities of dissident (Davenport 2007; Hibbs 1973; Muller 1985). Repression is relatively low in authoritarian regimes since the regimes have the capacity to eliminate potential challengers, which leaves the cost of protest against the regime relatively high. However, some empirical work (Regan and Henderson 2002) posits that semi-democratic or hybrid regimes are more prone to govern dissent with repression, since these regimes often face strong democratic pressure from people who may threaten their stability. Since Malaysia is regarded as a semi-democratic regime, I expect that repression in this country will appear to be intensifying because the government is highly likely to perceive dissent as a threat.

Some studies of how repression affects political participation (e.g. Gurr 1970; Hibbs 1973; Khawaja 1993) substantiates that repression has cumulative effects. Typically, state repression has a negative effect on dissidents (Hibbs 1973) because individuals are expected to participate in politics when the benefits are greater than the costs (Clarke *et al.* 2004; Riker and Ordeshook 1968). Therefore, any increase in costs may lead to dissident disintegration and constrain citizens from participating in collective actions. Simply, increased state repression is likely to decrease participation in any form. The deprivation theory, however, makes the opposite claim. As Gurr (1970) notes, state repression could generate 'collective frustration', which in turn escalates dissidents' interest in putting more pressure on the regime so dissenters can survive. From this perspective, repression can be seen to encourage dissent. Repression can also have both negative and positive effects. This is evidenced in the empirical findings of Bautista (2015) in Chile during the era of dictatorship (1973–1990). Bautista (2015) found that the repressed were more likely to participate in human rights organisations rather than being members of political parties or trade unions. This, it is argued, was the case since the decline in membership of political parties and unions was more pronounced amongst the repressed than non-repressed.

To analyse the impacts of the regime in Malaysia on youth political participation, my own argument aligns with Hibbs' view that repression discourages citizens' participation. In this case, I assume that state repression through implementation of new laws, amendments to existing criminal laws and violent repression inhibit young people in Malaysia from actively engaging in politics.

Understanding youth political participation

Citizen participation is crucial for a strong and effectively functioning democracy. Both participatory democrats (Barber 1984; Pateman 1970) and democratic realists (Sartori 1987; Schumpeter 1943) argue that citizen participation is an important condition for democracy, while differing about the degree to which it is desirable. Acknowledging that high levels of participation are a positive sign of democratic legitimacy (Lipset 1961; Norris 2002), representative institutions in democratic settings play a significant role in stimulating citizen participation by providing more participatory spaces and political collectivities (Young 2000). In this regard, we could expect that citizens in democratic regimes are more likely to engage in politics since there is less fear of repression and a more encouraging environment for participation.

However, over the past few decades, many established democracies have faced a 'crisis of liberal democracy' (Dalton 2008) or 'democratic deficit' (Norris 2001) due to the remarkable decline of political participation amongst young people with regards to the conventional, 'elite-directed' or institutionalised political participation, such as voting, party membership and trade unions. In this regard, there is a pessimistic account from many quantitative studies that strongly hold that young people are an apolitical generation because they participate at lower levels, including voting, organisational membership and civic life (Kimberlee 2002; Pirie and Worcester 1998). The decline of engagement in conventional politics in advanced democracies links to a number of recognisable explanations, including life-cycle effects or changing life patterns (Kimberlee 2002; Verba and Nie 1972), generational change (Bessant 2014; Grasso 2011, 2013, 2014, 2016; Grasso et al. 2017), the rise of 'new' issues and new social movements (Giugni and Grasso 2015; Grasso and Giugni 2013, 2016a, 2016b) and a growing cynicism about political systems and politicians (Henn et al. 2005) encouraging many young people to reject electoral politics (Grasso 2018; Pickard and Bessant 2017).

The situation is altogether different in newer democracies or semi-democratic regimes like the Ukraine. Numerous studies find young people in newer democracies are disengaged from politics, whether in conventional or unconventional forms of politics, compared to their counterparts in established democracies (Robertson 2009; Wallace and Kovatcheva 1998). The legacy of authoritarianism implied by a one-party system or significant levels of political corruption, as well as dissatisfaction about neo-liberal reforms, have all been identified as factors leading to political disengagement in newer democracies (Howard 2003).

Others, however, point to the way that more young people are turning to new informal, elite-challenging or non-institutionalised political participation like the rise of protest activism, online actions, political consumerism and social movements (Norris 2007; Zukin et al. 2006). These new forms of political participation, though, may be considered as 'less political' (Quintelier 2007: 167), but are more attractive to young people. Young people are attracted to these forms of participation because of characteristics such as them being loosely structured (Wuthnow 1998), ad hoc and focusing on a clear issue (Inglehart 1990). This optimistic view relies on the argument that many researchers have adopted an overly narrow conception of politics when understanding youth participation, without exploring how young people themselves define politics (O'Toole et al. 2003) and considering the connection between young people and contemporary politics (Bessant 2014). Heavy reliance on a very restrictive concept of participation could undermine the importance of other 'alternate' forms of politics which are often discounted as 'politics'. Only when a broad notion of politics is applied by researchers on youth participation would they find evidence of high levels of youth political activism (Henn *et al.* 2002). In other words, it is necessary to reconceptualise political participation to include new political repertoires that lie outside mainstream politics if we are to better understand why young people are not interested in politics, mainly in conventional politics.

I use a broad conception of politics by focusing not only on traditional participation but also the newer forms of participation in the following analyses. This framing might be used to understand how young Malaysians conceptualise politics and engage in political activities under a repressive regime, and how they resist government's effort to criminalise democratic dissents and political activity in Malaysia.

State repression and young people's political participation in Malaysia

As discussed above, several forms of political repression are imposed by the regime in Malaysia to limit young people's political freedom and restrain certain forms of political participation. What are the impacts of such efforts on youth political participation? Based on in-depth interviews with ten young non-participants, mainly those who have cast off politics by not voting or by disengaging from political and civic activities, we find that the main reason behind their political disengagement is strongly related to a 'culture of fear' due to the government's oppressive laws, such as the Sedition Act 1948 and the Universities and University Colleges Act 1971, which stifle political rights and freedoms. These laws, which mostly date back to the colonial rule era, give power to the regime to continue 'blocking' the opportunity for citizens' participation by subverting the democratic process.

All respondents stated that the increasing repressive measures taken by the regime to criminalise democratic dissent in Malaysia has clearly constrained them from actively engaging in politics more generally, as reflected in the comments by Nai[3] (26 years old):

> I think we are not completely free to participate in politics because there are still barriers that restrict our freedom. For example, students are not free to engage in political activities since the UUCA law has restricted their political rights.

In a similar vein, Munirah (24 years old) also argued that: 'I think the violent repression used by the government on dissidents scared not only me, but young people as whole, from getting involved in politics.' This study concluded that young people exposed to state-led legal force are more likely to be fearful citizens, and this deeply embedded culture of fear has created a very passive youth, who are, in fact, disengaged from both formal and informal political participation, including protest activism. In this regard, fear of government criminalisation of dissent resulted in young Malaysians perceiving political activities in Malaysia as high-cost or high-risk activities. Our findings are broadly consistent with previous literature on state repression (Davenport 2007; Hibbs 1973; Tilly 1978) that political repression increases the costs of collective action, which in turn may decrease individuals' desire to participate.

When asked how they define 'politics', most respondents (7 out of 10) tend to relate politics with the government and political elites, as argued by Satish (26 years old): 'Politics is about the government and it is an arena for people who want to show their leadership skills.' Since politics is framed only in terms of political actors, thus some (5 out of 10) viewed politics in a negative way by associating politics with political elites who are more concerned with pursuing power and self-interest rather than championing societal issues. This mirrors the findings of Henn and Foard (2014) and White *et al.* (2000) that young people in Britain are sceptical about politicians and considered them as self-serving elites.

Even worse, several respondents (3 out of 10) claimed that the word 'politics' itself is vague and complicated, as stated by Hidayah (24 years old): 'I know nothing about politics.' It could be seen, therefore, that these young non-participants who opted not to register as voters or become a member of any political organisation have a narrow conception of politics and regard politics as far removed from their daily lives. Indeed, an overly limited understanding of politics amongst young Malaysians may be one reason for their lack of interest and motivation to participate politically.

In addition, the government's excessive control over the electoral process, using the Electoral Commission as their primary agent, has created a system that is less than fair, particularly for opposition parties. The Electoral Commission, designed by a constitution to be an independent body, has in Malaysia been dogged by criticism due to serious malpractices – such as a large number of phantom voters in the electoral rolls, manipulation in delineating electoral con-stituencies, media bias towards the incumbent regime and a worsening practice of money politics (Lim 2002). The regime's manipulation of the electoral process appears to make many young people lose their trust and interest in par-ticipating in the political process, especially in the elections since they have been denied a fair vote. As argued by Khai (33 years old): 'I think sometimes I feel doubt about the voting system as it is not clearly transparent. When the system is not transparent, we cannot make much change through the election.' Azlin (32 years old) also claimed that: 'We can make a big change through voting. But it depends on the transparency and integrity of the electoral system and the vote itself. The more transparent the system, the more changes we can make.' This means that voting can bring changes to the system if the government ensures the transparency of the process and system. Since the established formal system is highly problematic and heavily control by the regime in Malaysia, more young people are turning to informal forms of participation. Based on the responses, all young non-participants do not fully abstain from engaging in political activities; instead, online activism and informal political discussion have been commonly used as direct political actions to replace traditional forms of participation. As commented by Rizal (32 years old): 'I think discussing political issues with friends and accessing social media are the only ways we can express our political views freely.' This is parallel to Meyer's (2004) argument that people turn to non-conventional means when they believe that conventional routes to influence are either unavailable or ineffective.

Although none of the interviewees have participated in protest activism, including demonstration or social movements, half of them (5 out of 10) believed that protest activism is part of democracy and that it is important to protest when we disagree with something or are discontented with the government. As Airen (25 years old) argued: 'Protest is a way of expressing our dissatisfaction on certain issues. We can demonstrate as long as it is organised in a proper way and not disturbing the public.' Azizul (34 years old) is also in agreement with this view: 'Protest is important to make our voices heard and to ensure the government pay attention to us.' Many young Malaysians feel grievances, anger and discontent about the government and the political system, but these emotions are not strong enough to motivate them to participate in protests because they have a shared belief that protesting is a risky activity that carries negative consequences to the protester. As popular mobilisation is deemed a threat to the government, the government claims that protest 'is not going to be good for us, and that is not an accepted culture in our country' (Najib 2016), and continues to criminalise dissent using legal and excessive force.

However, not all young people whose politics are repressed in Malaysia are afraid to resist the regime and demand changes. Despite the growing level of state repression limiting citizens' participation, a small group of young activists continue to channel their 'anger' and 'frustration' into political activism. This corresponds with the argument of Gurr (1970) that repression generates 'collective frustration' that might increase the likelihood of collective action. More than half of the young activists interviewed (6 out of 10) argued that they were driven to participate by their feeling of deprivation, injustice and indignation towards the repressive regime. As one activist, Syukri (32 years old), explained:

> we can see that our country is not being led in a better direction. We are moving towards a 'failed state' in terms of the economy, politics and society. So we can conclude that our country requires a total change, not a small-scale reform.

This opinion is also developed by Safwan (28 years old) in his comment: 'We are fighting to see Malaysia better than what it is now. This is not for us, but it is for the next generation.'

In Malaysia, there are several cases where social movements have been successful in changing policies. I refer, for example, to a national campaign for the abolition of the ISA spearheaded by a coalition of human rights NGOs, such as Aliran, Hakam and Suaram, under a banner called 'Anti-ISA Movement' (Gerakan Mansuh ISA, GMI) since 2001. After a decade of resisting and protesting against indefinite detention without trial, only in 2011 did the government repeal the Act and the detainees of ISA. Furthermore, nearly all youth activists (8 out of 10) claimed that they have been victimised and detained by the police at least once during their political activism, as reflected by Adam (25 years old): 'I was arrested more than ten times.' Despite this, most activists reported that they were not afraid, and the government's crackdown could not stop them struggling for a better future. They believed the experience of previous political figures was a source of inspiration for resisting the system.

There are several prominent forms of resistance, both violent and non-violent, that have been commonly used by youth activists against the repressive regime. First, protest activism, mainly street demonstration, is seen as a powerful tool long used by activists to challenge the status quo in Malaysia. However, large-scale protests only emerged in the 1990s, particularly during the 1998 *Reformasi* movement. Today, the most constant protest movement in Malaysia is a series of Bersih rallies that have been staged five times between 2007 and 2016. Malaysians' peaceful demonstrations normally descend into violence as the police impose brutality and excessive force against protesters. This evidence is consistent with the conclusions of Tarrow (1998) that protest in hybrid regimes tends to be more frequent and more violent because citizens do not have full access to political institutions.

Second, recent protests in Malaysia were more organised than previously due to the emergence of social movements. Social movements drive ordinary people to participate in politics, especially young people, because of their salient characteristics such as voluntary membership, nonpartisan, organised collective action, being less institutionalised and steadfast in promoting social changes. It is noteworthy that many social movements in Malaysia are closely linked with NGOs towards advocacy on issues of human rights and political freedom. These movements, as argued by Weiss and Saliha (2003), are deep-rooted in the middle class which embraces the idea of bringing about sociopolitical changes through large-scale protest.

Finally, the politics of popular culture is a relatively new, non-violent form of resistance used by youth activists to bring about change. The rise of popular culture in Malaysia can be seen during the *Reformasi* of 1998. Many cultural revolutionary groups emerged, created by a handful of young people such as Komunite Seni Jalan Telawi and Universiti Bangsar Utama. These were actively organised community-based programmes, educational street theatre and agitprop, not only challenging the authoritarian regime, but about being more embedded in everyday life politics. Today, this repertoire has expanded into different forms such as political jokes, satirical literature and internet memes, which are generally anti-establishment and non-ideological but aim to raise sociopolitical agendas. For instance, the protest movement in Malaysia, mainly Bersih, also utilised popular elements by staging a live concert and designing posters and images during the protest.

Conclusion

This chapter sheds light on the impacts of state repression on youth political participation in Malaysia by employing qualitative interviews drawn from 20 young Malaysians. From the findings, it appears that state-led repression has a negative effect on youth political engagement in Malaysia. It has created a 'culture of fear' which constitutes the main barrier to young people being politically active. Since participation in politics in this country is somewhat risky and involves such high cost, only a 'small circle' of young people who are brave enough and have a high level of commitment to democratic change are willing to challenge the status quo through social movements, protest and the politics of popular culture. The remainder of young people remain politically inactive or engage in low-risk channels like informal political discussion. Therefore, the large-scale exit of young people from politics makes the transition to democracy in Malaysia slow and difficult to achieve in the near future.

Note

The findings in this chapter are partially based upon the author's unpublished PhD thesis, Department of Politics, University of Sheffield, UK.

1 Malaysian young people are those aged 15–40 years old.
2 The UUCA was enacted in 1971 and aimed to provide a guideline for the establishment, regulation and administration of colleges and public universities. Following the student-led demonstrations in 1974, the government amended Section 15 by prohibiting any students or faculty members from expressing support for any political party, organisation or trade union. However, it was revised in 2012 to give greater flexibility to the students by allowing them to participate in political activities off-campus, including the right to be a member of any political party.
3 Names of interviewees have been changed to protect their identities.

References

Amnesty International. (2016). *Critical crackdown: Freedom of expression under attack in Malaysia*. London: Amnesty International Ltd.

Barber, B. (1984). *Strong democracy: participatory politics for a new age*. California: University of California Press.

Barraclough, S. (1985). The dynamics of coercion in the Malaysian political process. *Modern Asian Studies*, 19(4): 797–822.

Bautista, M. (2015). *Political effects of state-led repression: The Chilean case*. Available from www.barcelona-ipeg.eu/wp-content/uploads/2015/09/Bautista-Repression-Sep15.pdf [accessed 13 May 2017].

Bessant, J. (2014). *Democracy bytes: new media and new politics and generational change*. Basingstoke: Macmillan Palgrave.

Bessant, J. (2016). Democracy denied, youth participation and criminalising digital dissent. *Journal of Youth Studies*, 19(7): 921–937.

Clarke, H., Sanders, D., Stewart, M. and Whiteley, P. (2004). *Political choice in Britain*. Oxford: Oxford University Press.

Dalton, R. (2008). The quantity and the quality of party systems: party system polarization, its measurement and its consequences. *Comparative Political Studies*, 41: 899–920.

Davenport, C. (1995). Multi-dimensional threat perception and state repression: An inquiry into why states apply negative sanctions. *American Journal of Political Science*, 39(3): 683–713.

Davenport, C. (2007). *State repression and the domestic democratic peace*. Cambridge, UK: Cambridge University Press.

Davenport, C. (2009). Regimes, repertoires and state repression. *Swiss Political Science Review*, 15(2): 377–385.

Davenport, C. and Armstrong, D. (2004). Democracy and the violation of human rights: A statistical analysis from 1976 to 1996. *American Journal of Political Science*, 48(3): 538–554.

Earl, J. (2011). Political repression: iron fists, velvet gloves and diffuse control. *Annual Review of Sociology*, 37: 261–284.

Fernandez, L. (2008). *Policing dissent: social control and the anti-globalization movement*. London: Rutgers University Press.

Giugni, M. and Grasso, M. (2015). Environmental movements in advanced industrial democracies: heterogeneity, transformation, and institutionalization. *Annual Review of Environment and Resources*, 40: 337–361.

Goldstein, R. (1978). *Political repression in modern America: from 1870 to the present*. Cambridge, MA: Schenkman.

Grasso, M. (2011). *Political participation in Western Europe*. DPhil thesis, Nuffield College, University of Oxford.

Grasso, M. (2013). The differential impact of education on young people's political activism: comparing Italy and the United Kingdom. *Comparative Sociology*, 12(1): 1–30.

Grasso, M. (2014). Age-period-cohort analysis in a comparative context: political generations and political participation repertoires. *Electoral Studies*, 33: 63–76.

Grasso, M. (2016). *Generations, political participation and social change in Western Europe*. London: Routledge.

Grasso, M. (2018). Young people's political participation in times of crisis. In: S. Pickard and J. Bessant (eds), *Young people re-generating politics in times of crises*. London: Palgrave.

Grasso, M., Farrall, S., Gray, E., Hay, C. and Jennings, W. (2017). Thatcher's Children, Blair's Babies, political socialisation and trickle-down value-change: An age, period and cohort analysis. *British Journal of Political Science*, First View DOI: https://doi.org/10.1017/S0007123416000375.

Grasso, M. and Giugni, M. (2013). *Anti-austerity movements: old wine in new vessels?* XXVII Meeting of the Italian Political Science Association (SISP), University of Florence, Florence, 12–14 September.

Grasso, M. and Giugni, M. (2016a). Protest participation and economic crisis: The conditioning role of political opportunities. *European Journal of Political Research*, 55(4): 663–680.

Grasso, M. and Giugni, M. (2016b). Do issues matter? Anti-austerity protests' composition, values, and action repertoires compared. *Research in Social Movements, Conflicts and Change*, 39: 31–58.

Gurr, T. (1970). *Why men rebel*. Princeton: Princeton University Press.

Karim, H. and Hamid, S. (1984). *With the people: The Malaysian student movement 1967–74*. Petaling Jaya: Institut Analisa Sosial.

Henn, M. and Foard, N. (2014). Social differentiation in young people's political participation: The impact of social and educational factors on youth political engagement in Britain. *Journal of Youth Studies*, 17(3): 360–380.

Henn, M., Weinstein, M. and Forrest, S. (2005). Uninterested youth? Young people's attitudes towards party politics in Britain. *Political Studies*, 53: 556–578.

Henn, M., Weinstein, M. and Wring, D. (2002). A generation apart? Youth and political participation in Britain. *The British Journal of Politics & International Relations*, 4(2): 167–192.

Hibbs, D. (1973). *Mass political violence: A cross-national causal analysis*. New York: Wiley.

Howard, M. (2003). *The weakness of civil society in post-communist Europe*. Cambridge: Cambridge University Press.

Inglehart, R. (1990). *Culture shift in advanced industrial society*. Princeton: Princeton University Press.

Kimberlee, R. (2002). Why don't young people vote at general elections? *Journal of Youth Studies*, 5(1): 85–97.

Khawaja, M. (1993). Repression and popular collective action: evidence from the West Bank. *Sociological Forum*, 8(1): 47–71.

Lim, H. (2002). Electoral politics in Malaysia: managing elections in a plural society. In: A. Croissant, G. Bruns and M. John (eds), *Electoral politics in Southeast and East Asia*. Singapore: Friedrich Ebert Stiftung.

Lipset, S. (1961). *Political man: The social bases of politics*. Baltimore: Johns Hopkins University Press.

Lucan Way. (2014). The lessons of 1989. In: L. Diamond and M. Plattner (eds), *Democratization and authoritarianism in the Arab world*. Baltimore: Johns Hopkins University Press.

Meyer, D. (2004). Protest and political opportunities. *Annual Review of Sociology*, 30: 125–145.

Muller, E. (1985). Income inequality, regime repressiveness and political violence. *American Sociological Review*, 50: 47–61.

Najib, T. (2016). 'Street protests not going to make me bow out, says PM'. *The Star*, 18 November. Available from www.thestar.com.my/news/nation/2016/11/18/najib-i-will-not-step-down-street-protests-not-going-to-make-me-bow-out-says-pm/ [accessed 17 May 2017].

Norris, P. (2001). *Digital divide: civic engagement, information poverty, and the internet worldwide*. Cambridge: Cambridge University Press.

Norris, P. (2002). *Democratic phoenix: reinventing political activism*. Cambridge: Cambridge University Press.

Norris, P. (2007). Political activism: new challenges, new opportunities. In: C. Boix and S. Stokes (eds), *The Oxford handbook of comparative politics*. Oxford: Oxford University Press.

O'Toole, T., Lister, M., Marsh, D., Jones, S. and McDonagh, A. (2003). Tuning out or left out? Participation and non-participation among young people. *Contemporary Politics*, 9(1): 45–61.

Pateman, C. (1970). *Participation and democratic theory*. Cambridge: Cambridge University Press.

Pickard, S. and Bessant, J. (eds). (2017). *Young people re-generating politics in times of crises*. London: Palgrave Macmillan.

Pirie, M. and Worcester, R. (1998). *The millennium generation*. London: Adam Smith Institute.

Quintelier, E. (2007). Differences in participation between young and old people. *Contemporary politics*, 13(2): 165–180.

Regan, P. and Henderson, E. (2002). Democracy, threats and political repression in developing countries: are democracies internally less violent? *Third World Q*, 23(1): 119–136.

Riker, W. and Ordeshook, P. (1968). A theory of the calculus of voting. *American Political Science Review*, 62: 25–42.

Robertson, F. (2009). *A study of youth political participation in Poland and Romania*. PhD thesis, University College London. Available from http://discovery.ucl.ac.uk/18725/ [accessed 2 May 2017].

Sartori, G. (1987). *The theory of democracy revisited*. Chatham, NJ: Chatham House Publishers.

Schumpeter, J. (1943). *Capitalism, socialism and democracy*. London: Allen & Unwin.

Snyder, D. (1976). Theoretical and methodological problems in the analysis of governmental coercion and collective violence. *Journal of Political and Military Sociology*, 4: 277–293.

Suhaimi, A. (2013). *Politik mahasiswa sudah mula hilang arah* [online]. The Malaysian Insider. Available from www.themalaysianinsider.com/Malaysia.article/politik-mahasiswa-sudah-mula-hilang-arah/ [accessed 11 May 2017].

Tarrow, S. (1998). *Power in movement: social movements and contentious politics*. Cambridge: Cambridge University Press.

Tilly, C. (1978). *From mobilization to revolution*. Reading: Addison-Wesley.

Verba, S. and Nie, N. (1972). *Participation in America: political democracy and social equality*. New York: Harper & Row.

Wallace, C. and Kovatcheva, S. (1998). *Youth in society: The construction and deconstruction of youth in East and West Europe*. New York: Macmillan Press.

Weiss, M. (2006). *Protest and possibilities: civil society and coalitions for political change in Malaysia*. California: Stanford University Press.

Weiss, M. (2011). *Student activism in Malaysia: crucible, mirror, sideshow*. Ithaca; Singapore: Cornell Southeast Asia Program and NUS Press.

Weiss, M. and Saliha, H. (2003). *Social movement in Malaysia*. London: Routledge Curzon Ltd.

White, C., Bruce, S. and Ritchie, J. (2000). *Young people's politics: political interest and engagement amongst 14–24 year olds*. York: Joseph Rowntree Foundation.

Wuthnow, R. (1998). *Loose connections: joining together in America's fragmented communities*. Cambridge: Harvard University Press.

Young, I. (2000). *Inclusion and democracy*. Oxford: Oxford University Press.

Zukin, C., Keeter, S., Andolina, M., Jenkins, K. and Delli Carpini, M. (2006). *A new engagement? Political participation, civic life, and the changing American citizen*. New York: Oxford University Press.

Chapter 12

Russian politics of radicalisation and surveillance

Anna Schwenck

The criminalisation of dissent and the production of legal uncertainty are mainstays of authoritarian political figurations.[1] Compared with most cases in this volume, this chapter on Russian politics of radicalisation constitutes an extreme case (Gerring 2006: 101). As such, it offers a better understanding of the logics of state surveillance practices and state responses to dissent.

I make both a general and a specific argument in this chapter. My general argument concerns an altered usage of the concept of radicalisation. With the increase in large-scale violent attacks in Euro-American societies, counter-radicalisation strategies, including surveillance, have multiplied. These strategies draw on specific conceptualisations of radicalisation (Busher and Macklin 2015). Most literature reads radicalisation as a process which pertains to non-state actors. Hence, the actors who are assumed to become radicalised are individuals and groups. Although the relative nature of radicalisation is widely acknowledged (Sedgwick 2010), the state jurisdiction has remained a static benchmark to trace radicalisation pathways. Instead, radicalisation will be understood here as a process through which a *state* radicalises. I define the state as a relation between a polity and a people, not as a discrete entity. Therefore, the actions and narratives of civil society actors and academic experts are co-constitutive of state radicalisation and surveillance practices. Radicalisation thus means the gradual increase in surveillance and control practices within a polity. As a working definition, de facto granted personal freedoms, due process, and legal certainty serve as benchmarks to assess state radicalisation. When state actors declare narratives and actions to be extremist, they portray them as being above political contestation (Buzan *et al.* 1998: 24). This often means that acts of surveillance, i.e. practices of identification, monitoring and analysis which the state can use to enforce social control, are presented as inevitable to guarantee national and personal security (Monahan 2010: 10, 102). When states radicalise, the spectrum of sanctioned political debate and civic engagement narrows.

My specific argument is that state youth politics and anti-extremist policies in Russia, both of which have considerably expanded since the early 2000s, share a common ideological core. To date, scholars of state youth politics

(Lassila 2014; Hemment 2015) and students of anti-extremist policies (Epshtein and Vasil'ev 2011; Kravchenko 2015; Gaufman 2017) have taken little note of the ideological overlaps between the two policy fields. First, both are informed by the organic nationalist axiom that criticism of the government either results from foreign insinuation, or indicates an immature national consciousness. The two policy fields can thus be understood as complementary in safeguarding a particular image of the nation and state. While anti-extremist policies are designed to ward off putative pestilent external influences from the body politic, youth policies aim to strengthen the body politic from within by fostering government-loyal identifications. Thus, both policy fields attempt to watch over 'the young generation' with the stated aim to surveil, protect and guide it (Lyon 2009).

Second, both policy fields feed on the background beliefs that youthfulness is a transition state from immaturity to maturity and that 'the young generation' constitutes a risk group, inclined to extreme views. Consequently, young adults are understood as vulnerable to so-called extremist influences and as needing guidance by experts and the state (Omel'chenko 2006: 156). Finally, circulating threat scenarios which have the demise of the state and nation as their telos lend the two policy fields urgency. These threat scenarios depend on the public framing of particular phenomena as national threats. Youth politics and anti-extremist policies can thus be understood as different answers to the same threat scenarios.

In the following, I will first elaborate on the notion of 'threat frame' as a way to study alleged threats to national security. Then, I will analyse two threat frames that were crucial to the radicalisation of the Russian state. I call them the Orange and the Pink Threat. While the 'Orange Threat' legitimised youth political and anti-extremist measures that would hamper dissent in the aftermath of the Ukrainian Orange Revolution, the 'Pink Threat' classified LGBT and feminist social activities as 'extremist', i.e. as jeopardising the Russian state and nation. In both cases, youth political and anti-extremist measures complemented each other in obstructing dissent and intensifying state surveillance.

National threat frames

Political scientist Elizaveta Gaufman studies security threats by analysing so-called enemy images [*Feindbilder*]. She defines an enemy image as 'a personified existential threat that is embedded in collective memory' (Gaufman 2017: 65). I propose a revised model to study threats to national security. It builds on Gaufman's refinement of the securitisation approach, but departs from it in key aspects. The study of enemy images is rooted in the transposition of findings about an individual's psyche to the group level. The framing approach, in contrast, assigns primacy to the social. It defines frames as the result of interaction – as schemata of interpretation that provide definitions of a given situation (Goffman 1986: 11 and 21).

The framing approach is productive for the study of threat scenarios in authoritarian figurations. First, under authoritarian rule, it is harder to reveal fabrications publicly, and researchers are more often compelled to doubt official proclamations. With regard to Russian domestic politics, there is an implicit scholarly debate about whether government officials *really believed* in a threat or just *fabricated* it as a justificatory pretext (Duncan 2013). The framing approach solves the question of whether a threat scenario was instrumentally constructed or emerged due to the actors' taken-for-granted beliefs on a theoretical level. It combines an instrumentalist perspective with a cultural-interactionist one: 'While actors instrumentally frame situations so as to press their case, their very understanding of what is instrumental is shaped by taken-for-granted frames' (Polletta and Ho 2006: 188).

Second, comparatively little research deals with *societal* processes in authoritarian figurations. The framing approach compels the researcher to take society in account, because the resonance of a frame, i.e. its capacity to effectively shape citizens' perceptions, depends on its connection to existing public narratives: '[…] stories precede frames, stories make frames compelling, and stories overshadow frames in mobilising power and as a political resource' (Davis 2002: 25). Thus, public narratives plough the soil in which a frame can take root. Public narratives are stories that circulate through their telling and retelling in the most varied settings, from family gatherings to international media (Somers and Gibson 1993). Finally, discussion of public narratives (instead of collective memory) emphasises the transnational dimension of national threat frames. Public narratives can be shared, *although* their tellers might strongly identify with different group histories. The national threat frames analysed below are in several aspects specific to Russia, but are nevertheless often supported by narratives that are not. This aspect underlines that authoritarian figurations are also stabilised by narratives and actions that circulate across state borders.

National threat frames become meaningful in relation to what I call the 'security contract of the nation-state'. According to Mabel Berezin, the provision of security is a central function of nation-states: 'The modern nation-state is the institutional location of a relation between a polity and a people that provides security for its members' (Berezin 2009: 6). Understanding the nation-state as the location of a *relationship* emphasises the active participation of people – elites and non-elites – in its 'making' (Gal and Kligman 2000: 20). Consequently, a nation-state's provision of security takes the form of a contract between its members. The contract is based on mutual solidarity. From this perspective, giving birth or serving the armed forces amount to individual trade-ins for collective security.

When national threat frames resonate, they tighten the conditionality already in place. Citizens who hold ambiguous views during politics as usual might feel compelled to force a particular ideological line when threats seem to be approaching (Swidler 1995: 35). The more a situation is charged with emotional concern for the nation, the greater the chance that non-compliance counts as unpatriotic behaviour and that surveillance appears to be necessary (Monahan 2010: 3–4). National threat frames are socially relevant, because they have real effects on everyday interactions. They may expedite the criminalisation of dissent, widen the spectrum of suspicious behaviour and increase public demand for surveillance.

Anthropologist Mary Douglas' reflections on the relationship of purity and danger indicate two crucial reference points for the following analysis of national threat frames. She argues that both the perceived pressure on a group's external boundaries and the transgression of its internal lines (its structure of domination) may be perceived as danger and arouse notions of purity (Douglas 2005: 157). If we understand the moral panics of modern societies as secular concerns about purity, they are indicative of threat frames. Moral panics therefore constitute a pivotal heuristic to detect not only national threat frames, but also public narratives that subtly stabilise these frames and make them compelling. Moreover, concerns over external borders are often justified with reference to internal lines, and vice versa. For instance, the framing of certain behaviour as a national threat is often morally reinforced through stories that cast this behaviour as alien. Such stories can be understood as moral backings of national threat frames. They convey that alien behaviour has permeated the external boundary. They thereby portray the domestic structure of domination as an object of cultural value (e.g. as a unique national tradition) that is to be protected. Thus, stories that provide moral backing may depict domestic dissent as alien, non-natural behaviour.

The Orange Threat

In the winter of 2004, young adults were amongst the most active protesters who took to the streets in Ukraine after evidence about electoral fraud became public (Stepanenko 2005: 35–36). The protesters marched mostly under orange banners, the campaign colour of Viktor Iushchenko, who was the main competitor of the Kremlin-endorsed presidential candidate. Their mobilisation led to the inauguration of the EU and NATO-friendly Iushchenko in January 2005.

While this outcome was celebrated as an 'Orange Revolution' by Central European and Northern American media outlets, Russian government officials portrayed it as the result of a planned strategy by NATO governments and maintained that Western governments had trained *radical* youth groups. They assumed that a similar scenario might happen in Russia and thwart Vladimir Putin's re-election in 2008, and portrayed Western-funded trainings for NGOs and youth groups as an insidious attack on Russian sovereignty (Horvath 2013: 90).

This trope of 'malign outside intervention' is a structural characteristic of the Orange Threat frame. The trope refers to the personalistic, anti-capitalist suspicion that behind every phenomenon, a financial tycoon maybe be hiding. It juxtaposes open conflict with intrigue and doublespeak, often associated with both Jewishness and femininity (Mosse 2010). The trope therefore leaves ample room for anti-Semitic and misogynistic ascriptions. Moreover, the trope pertains to an organic nationalism which denies domestic cleavages and transposes them to the international arena (Mann 2004: 63). Because organic nationalism defines all dissent as a result of foreign insinuation, dissenting citizens are not treated as political opponents, but as national traitors who lack patriotic feelings (Morozov 2015: 151). Domestic political conflicts appear to be nothing more than reflections of conflicts between great powers. The widely used term 'geopolitics' lends this organic nationalist reasoning a scientific frame of reference. It gives those who frequently refer to geopolitics the semblance of rationally discriminating actors (cf. Adorno *et al.* 1950: 621–623).

The plot of outside intervention received moral backing from narratives about the collaboration of Ukrainian nationalists with German Nazis during the Great Patriotic War.[2] These narratives are rooted in Soviet discourse, which had simplistically portrayed Ukrainian nationalists as agents of German fascism (Rossoliński-Liebe 2014: 365).[3] Since 2000, the Russian government has largely appropriated the Soviet Union's mythologisation of the Red Army's fight against fascism as well as the functionalist usage of the Soviet anti-fascist terminology (cf. Bikbov 2014: 176). It has also increasingly justified its political course on the basis of a moral superiority due to the polity's 'historical' anti-fascist stance (Morozov 2015: 101). Correspondingly, political opponents are often accused of co-operating with fascists.

This was also the case during the Ukrainian presidential race, when supporters of the Kremlin-endorsed candidate overemphasised ideological overlaps between the orange coalition, the Ukrainian Far Right and historical Ukrainian nationalism (Kuzio 2005). While most government officials did not identify supporters of Iushchenko as fascists verbatim (Gaufman 2017: 115), their framing of the events left ample room for open nationalists to connect the dots in media coverage (Horvath 2013: 34).[4] Similarly, the government-sponsored group *Nashi* draws implicit connections between the Orange coalition and fascism in its choice of name. The group employs the self-designation 'youth democratic *anti-fascist* movement' (Schwenck 2015a). The term *nashi* (literally: 'the ours') was an informal expression for Red Army troops fighting against Nazi Germany, which signalled the speaker's pro-Soviet stance. Today, *nashi* invokes the dualism of the war time: those young adults who do not fight on 'our', i.e. the government's, side are traitors. While such a Manichean logic has been a recurrent pattern of Russian politics since the early 1990s (Urban 1994), its projection onto society and the younger generation marked the revival of so-called state youth politics.

The revival of state youth politics and the expansion of anti-extremism

The Orange Revolution served indeed as a catalyst for the generation of government-critical youth organisations in Russia (Sperling 2015: 83). The government responded to this manifestation of the 'Orange Threat' with the organisation of pro-government rallies and the sponsoring of anti-Orange youth groups. Some of these youth groups were newly founded. Others, such as *Nashi* and the *Young Guard of United Russia*, had been restructured and rebranded (Lassila 2014; Schwenck 2015b). Besides displaying support for the anti-Orange coalition, these youth groups were also sponsored to intimidate government-critical voices (Horvath 2013: 191).

The government's investment in rallies and youth groups was part of a more general package of measures that aimed to inhibit and criminalise dissent. Freedom of assembly was restricted by a 2004 law that gave authorities the power to ban demonstrations on the grounds of public safety (Johnson 2011). In 2007, the special police force (OMON) started to use tactics against protesters that had initially been developed to contain football hooligans who turned violent (Horvath 2013: 172). Moreover, the government pressed ahead with several legal and institutional propositions that were to control civic life. In 2006, an NGO law introduced a complicated registration procedure. Registration became dependent on the goals of an NGO, which should not threaten the Russian Federation's national interest, its national unity or cultural heritage – notions which the law did not define further (Pitts and Ovsyannikova 2015: 120). Moreover, so-called civic chambers, public consultative bodies, proliferated. While these institutions did foster participation in general, the right of the authorities to choose civil society representatives at their convenience facilitated the exclusion of dissenting voices (Richter 2009: 17). Finally, the government expanded the notion of 'extremist offences', which led to the persecution of liberal protesters and long-standing anti-fascist activists (Horvath 2013: 185–186 and 263).

The first legal reference to anti-extremism dates back to the June 2001 ratification of the Shanghai Convention on combating terrorism, separatism and extremism. In 2002, a federal law on fighting extremist activities shifted the focus from terrorist attacks to the *incitement* of racial, national, religious or social hatred in connection to violence. It amended paragraph 282 of the Criminal Code which, since 1996, defined the incitement of national, racial or religious hatred as an offence (Epshtein and Vasil'ev 2011: 149). Despite substantial criticism of the vague definition of extremism in the existing legislation, it was further tightened. In 2006, the condition of violence was omitted altogether. In 2008, a department to fight extremism with branches in all administrative districts was created (Epshtein and Vasil'ev 2011: 20–23). These and further legal measures were implemented against the backdrop of large-scale violent attacks against civilians, including the Moscow theatre hostage crisis (2002) and 'Russia's 9/11', the Beslan school siege, during which more than 180 schoolchildren died. The attacks left their mark on daily life in Russian cities. Since then, an ambient air of threat fills public spaces, exacerbated by continuous warnings to watch out for suspicious items, random security checks and intense racial profiling (Iuristy za konstitutsionnye prava i svobody [Jurists for constitutional rights and freedoms (an NGO)] 2006).

Most accounts of the revival of state youth politics in the early 2000s attribute the government's 'discovery' of young people as a critical political force to the Orange Threat. Youth sociologist Elena Omel'chenko connects the revival also to the media coverage about the 2005 car burnings in Parisian suburbs (for which 'Muslim migrant youth' was said to be responsible). Her remark highlights the transnational dimension of the moral panic about 'young adults' radicalisation' at the time. Moreover, she mentions the previous linkages (made by Russian academics and experts) between 'youth' and 'extremism' in the context of right-wing violence (Omel'chenko 2006: 154). Her view helps to account for the great variety of concerns that both youth politics and anti-extremist policies aim to combat. Equally, sociologist Peter Meylakhs' research indicates that the discovery of youth as a political group condensed several controversial issues into a threat relating to youth and the future of the nation. Building on concerns over high mortality rates and sexually transmitted diseases, debates about drug use, sex education, homosexuality and pornography dominated Russian media coverage during the late 1990s. Meylakhs has argued that these problem frames figured as 'symbolic polluters' of society for conservative activists, including Parental Committees and religious groups. This framing was largely reproduced in the media at the time. The main *object* of concern were children, whose symbolic purity and innocence was to be protected. It is important to note that 'children' stood for several age cohorts in such debates (Meylakhs 2009).

Several measures of state youth politics reflect this variety of concerns. For example, besides fostering young entrepreneurship, All-Russian Youth Educational Fora mainly promote conservative civic engagement, ranging from patriotic remembrance over volunteering for the socially disadvantaged to vigilante groups which surveil public space (Hemment 2015). While such fora have a special antiterrorism focus in the Caucasus regions (Comai 2012), they aim to provide young adults with greater opportunities for upward mobility and encourage them to become engaged in tune with the political course of the government. Conservative civic engagement is also promoted by organs of student self-government, the Movement of Youth Parliaments and state-run youth clubs (Kokorev *et al.* 2006: 107–108). All these measures were also publicly justified as remedies for diverse social problems, ranging from youth criminality over declining birth rates to morally reprehensible behaviour.

The same broad range of issues are also targeted by anti-extremist policies. Extremism is associated with terrorism, the state's disintegration (separatism and demographic issues) and the constitutive people's moral decline – i.e. all things considered to deviate from a perceived normality (Epshtein and Vasil'ev 2011: 24–27).[5] Such ambiguity is also supported by academic and expert publications about youth extremism. For instance, one publication considers radical Islamists, Russian skinheads as well as nationalists of Russia's 'traditional' ethnic minorities [*natsional'nosti*] to be extremists. All these extremists are portrayed as posing an equal threat to the future of Russia alongside sexually transmitted diseases, drug use and migration from poorer countries in the post-Soviet region (Surovtsev 2013). In a similar vein, a study that was co-conducted by a member of the Russian Academy of Sciences regards respondents' readiness to join *any* protest action to be a 'manifestation of extremist values' (Chuprov and Zubok 2009: 74–75).[6] All these accounts draw on the image of the vulnerable young adult who is inherently inclined to extreme views and actions. They assume that adolescents' extreme views and actions can easily be transformed into 'extremist' ones. Some experts believe that adolescents can easily be inveigled into deviant behaviour due to their quest for emancipation and their developing sexuality (Ekho Moskvy 2009). Anti-extremist prophylaxis should therefore primarily be aimed at the young generation. The Strategy for State Youth Politics emphasises that 'young people should be prepared for resisting manipulations and extremist appeals' (Government of the Russian Federation 2006).

The fight against extremism is directed less at hate crimes, the perpetrators of which often go unpunished (Epshtein *et al.* 2012: 114). Its main target is the *visibility* of phenomena that are framed as symbolic pollutants. For instance, anti-extremist law enforcement mainly targets the display and description of drug use (its 'propaganda'). The objects it surveils are publishing houses, public libraries and other distributors of cultural artefacts (Epshtein and Vasil'ev 2011: 127–148). Such surveillance and persecution of the public *visibility* of perceived deviance is also a structural characteristic of the Pink Threat.

The Pink Threat

Beginning in 2006, regional governments started to re-criminalise the display of same-sex relationships in Russia, peaking in a 2012 ban on so-called homosexual propaganda in the country's most liberal city, St. Petersburg.[7] The ban was followed by a national law in 2013 that made the distribution of positive information about so-called non-traditional intimate relationships amongst minors a punishable offence. The law was an amendment to the Youth Protection Act, which was introduced in 2012 to safeguard minors from the display of violence, drug consumption (including alcohol and nicotine) and 'values that would negate the nuclear family' (Wiebe 2013: 99). Thus, the figure of the vulnerable young adult is also a structural feature of the Pink Threat.

The Pink Threat shares another structural feature with the Orange Threat: the trope of malign outside intervention. Public narratives that support the Pink Threat frame often recount the foreign origins of both feminism and the LGBT movement. Queer and feminist ideologies are portrayed as having being implanted in the 1990s through Western aid schemes in order to weaken Russian sovereignty (Temkina and Zdravomyslova 2014: 260).

I will thus only focus on those structural characteristics that are specific to the Pink Threat, such as stories of pollution and narratives of the gender binary, which reiterate essentialised differences between women and men. Both genres of stories are disseminated, albeit in a different register, by institutions as different as the Russian Orthodox Church and make-over TV formats akin to *What not to Wear* (McRobbie 2009: 128; Lerner and Zbenovich 2013).

My focus on such stories is informed by the assumption that public anxieties about the body and sexual taboos correspond to anxieties about the external boundaries and internal lines of a social figuration (Douglas 2005: 152). Norms of sexual behaviour are not only surveilled to ensure national reproduction (Gaufman 2017: 146); their surveillance also aims to stabilise fundamental structures of domination and the rules by which political legitimacy is created (Gal and Kligman 2000: 28 and 34).

Stories of the heterosexual gender binary reiterate essentialised differences between the sexes and thereby justify particular social arrangements (Goffman 1977: 302). In the context of organic nationalism, they also safeguard the nation-state's security contract, which builds on the analogy between the heterosexual family and the nation (Riabov and Riabova 2014: 25). Any attempts to de-essentialise the idea of two discrete sexes questions the validity of notions such as the fatherland or the mother tongue (Riabov and Riabova 2014: 25). This is why patriarchy is central to the reproduction of nationalism, and vice versa (Gal and Kligman 2000: 25). Patriarchy constitutes 'a social, cultural, and political system that values men and the attributes of masculinity over women and the feminine qualities stereotypically affiliated with them' (Sperling 2015: 17). As traditional masculinity, it is predicated upon the downgrading of femininity (Pohl 2004). Therefore, the display of same-sex relationships as relationships between equal partners (which symbolises the equality of 'female' and 'male' sexuality) is as threatening to patriarchy as behaviour by women and men that does not fit classic attributions. The Union of Orthodox Women listed such classic attributions in their public response to Pussy Riot's 2012 punk prayer. They include 'tenderness, mildness, elegance, [...] natural wisdom, sensitiveness to beauty, self-sacrifice' (POLIT.RU 2012). According to an analysis of this statement by Anna Temkina, the Union ultimately rejected feminism because it 'threatens the patriarchal heritage, the "gender stereotypes of male culture"' (POLIT.RU 2012). Patriarchy is portrayed as an object of cultural value that is to be protected.

The Pussy Riot case shows that the dominance of the 'stereotype of true manliness' depends on its unquestioned public display (Mosse 2010: 5–7). Only when boundaries between masculinity and femininity are publicly called into question do they pose a threat to patriarchy. Remarks such as 'Sexual relationships are intimate, they should not be showcased' have been standard amongst those citizens who have opposed gay parades in Russian cities since the mid-2000s: here, already, the main point of criticism is the *visibility* of same-sex partnerships (Kon 2013: 60–61). A statement by a deputy of the Ukrainian Communist Party has a similar reasoning at its core (a similar ban was drafted in Ukraine in 2012): 'A kid [...] sees a same-sex couple imposing their relationship on him. That cannot be allowed. Children should be protected against such information' (as cited in Bershidsky 2012). This quote shows the degree to which public visibility alone is interpreted as potential seduction.

The idea that visibility is seductive is predicated on normative notions of sexual maturation. Young people are supposed to mature into heterosexual adults with a 'proud gender identity' – in analogy to the New Right idea of 'proud cultural identities' (Spektorowski 2003). It might be argued that it is hence the figure of the pederast which serves as the personification of the Pink Threat, and less the figure of the paedophile (Gaufman 2017: 147). The pederast might threaten the formation of a heterosexual adult identity, because he hinders the adolescent from learning how to control homosexual desire (McRobbie 2009: 107). Minors thus appear in need of protection, because neither their gender identity nor their sense of a moral sexual responsibility are fully developed. They are not able to control their desires, an idea that corresponds with the widespread notion that homosexuality results from loose morals (Sperling 2015: 73). On a fundamental level, such reasoning acknowledges that homosexual desire is normal (even natural) – why else should it be seductive for minors even to witness same-sex love?

It is therefore no surprise that the visibility of same-sex love became the target of state surveillance. In 2012, LGBT activists were amongst the groups that were most frequently accused of extremist activities (Kravchenko 2013). For instance, slogans displayed at an authorised feminist event in support of Pussy Riot were found by police to be extremist. They had to be removed together with a rainbow flag (Sperling 2015: 260). In 2013, Pussy Riot's Punk Prayer video was included in the Federal List of Extremist Material. In the same year, the head of a centre for perinatal care was accused of inciting hatred against men as a social group. The accusation warned that the centre would teach women to reject the family as a social institution and to act negatively towards their male spouses (Kravchenko 2014). It should be emphasised that the misuse of anti-extremist legislation does not only occur top-down – from the government to oppositional groups. The vague language of the legislation allows its misuse in economic and personal conflicts and in boosting law enforcement statistics (Kravchenko 2015). These caveats notwithstanding, the very fact that the head of a perinatal centre was accused of 'rejecting the family' exemplifies the social dimension of the Pink Threat.

Youth political legislation invokes the themes of nationalism, masculinity and pollution by referring to young citizens' morality and physical health on the one hand, and to a strong Russia on the other. State youth politics aim to increase the number of young adults who 'share universal and national moral values, are physically healthy, [...] love their fatherland and are ready to defend its interests and endeavour to dynamically develop a strong and independent Russian Federation' (Government of the Russian Federation 2014: 5). They support those young couples who are in legal marriages (i.e. heterosexual) and who are prepared to give birth to several children. The children's upbringing should be guided by 'traditional Russian values', the content of these, however, is not officially defined (Government of the Russian Federation 2014: 11).

Besides stories of pollution, public narratives which recount the gender binary permeate daily life in Russia. These narratives are not specific to Russia, but they found special support during the 1990s, when liberals envisioned women returning to their 'traditional' roles in a market society. Such visions appeared against the backdrop of the Soviet model of patriarchy which featured working mothers, female cosmonauts and snipers (Temkina and Zdravomyslova 2014: 259). The 1990s also brought an influx of the beauty products and services of style and self-help professionals that allowed for the democratisation of the 'endless stagings of femininity' (McRobbie 2009: 111) that had hitherto been a privilege of media stars. Today, TV shows reiterate stories which convey that female appearance and behaviour are the crux for finding a male partner and for keeping a husband (Lerner and Zbenovich 2013: 843–844). Such shows also indicate that in the last instance, 'true' femininity is predicated upon masculine complementarity: 'No one adorns a woman better than a properly matched man' (as cited in Lerner and Zbenovich 2013: 843).

Reiterating norms of sexual behaviour and gender complementarity constitutes a form of societal surveillance. Stories of pollution substantiate the assumption that the public display of 'non-traditional' sexuality interferes with heterosexual maturation. Stories of gender complementarity enshrine misogyny in everyday life and stabilise dominant patterns of legitimacy generation. Especially in the absence of regular democratic elections, masculinisation is a key strategy to gain political legitimacy (Sperling 2015). As such, these stories, which are structural characteristics of the Pink Threat, secure political legitimisation strategies and the prevalent structure of domination in everyday encounters.

Conclusion

The Orange Threat and the Pink Threat share two structural characteristics: the trope of malign outside intervention and the figure of the vulnerable young adult. Whereas the Orange Threat has been mainly stabilised by the reactivation of public stories about Ukrainian nationalism, the Pink Threat is stabilised by stories of symbolic pollution and, less explicitly, by narratives of the gender binary. In general, the radicalisation of the Russian state has been accompanied by the growth of anti-extremist and youth policies, and an increase in practices of surveillance. While anti-extremist policies monitor young adults' (sexual) behaviour, their appearance and the cultural products they can access, youth policies project the ideal of the young patriotic and heterosexual citizen. This ideal-type constitutes the foil for evaluations of deviant behaviour and the identification of extremist material, assumed to lead vulnerable youth and young adults astray.

These Russian politics of radicalisation and surveillance have been supported by both conservative civil society organisations and academic experts. I have tried to accentuate the societal pillars of the Russian law and order policies as well as citizens' share in surveillance practices – especially with regard to the norms of the gender binary. Given that the Russian political figuration is an authoritarian one, in which the state can more easily suppress dissent, the case at hand raises the question of citizen groups' and academic experts' role in the radicalisation of democratic states.

Notes

All translations from Russian are made by the author. All Russian terms and titles are transliterated from Cyrillic according to the modified Library of Congress system of transliteration without diacritics.

1 Norbert Elias (1978) rejected 'system' due to the concept's static connotations. 'Figuration' emphasises people's interdependencies. The notion can be applied to different types of solidaristic groupings (villages, state-societies, tribes).
2 The Great Patriotic War narrative starts with the German invasion of the Soviet Union in 1941. It excludes the Molotov-Ribbentrop pact.
3 This equation should also support the promotion of Soviet patriotism in the Ukrainian Soviet Socialist Republic (Rossoliński-Liebe 2014: 369).
4 Cf. for instance, 'The Orangers – carriers of neo-fascism' (Voitsekhovskii and Tkachenko 2008).
5 This is not specific to the Russian case, see Oppenhäuser 2011.
6 Both publications were supported by the Russian Ministry of the Interior.
7 For an overview of the status of same-sex relationships during Soviet times, see Sperling, 72.

References

Adorno, T., Frenkel-Brunswik, E., Levinson, D. and Sanford, N. (1950). *The authoritarian personality*. New York: Harper & Row.
Berezin, M. (2009). *Illiberal politics in neoliberal times: culture, security and populism in the new Europe*. Cambridge: Cambridge University Press.
Bershidsky, L. (2012). 'Curing' homosexuality in Russia and Ukraine. *Bloomberg*, 14 October. Available from www.democraticunderground.com/113720833 [accessed 7 March 2017].
Bikbov, A. (2014). *Grammatika poriadka. Istoricheskaia sotsiologiia poniatii, kotorye meniaiut nashu real'nost'*. Moscow: Izdatel'skii dom Vysshei shkoly ekonomiki.
Busher, J. and Macklin, G. (2015). Interpreting 'cumulative extremism': six proposals for enhancing conceptual clarity. *Terrorism and Political Violence*, 27(5): 884–905.
Buzan, B., Wæver, O. and Wilde, J. de. (1998). *Security: A new framework for analysis*. Boulder: Lynne Rienner Publishers.
Chuprov, V. and Zubok, I. (2009). *Molodezhnyi ekstremizm: Sushchnost', formy proiavleniia, tendentsi*. Moscow: Rossiiskaia akademiia nauk. Institut sotsial'no-politicheskikh issledovanii.
Comai, G. (2012). Youth camps in post-Soviet Russia and the Northern Caucasus: The cases of Seliger and Mashuk 2010. *Anthropology of East Europe Review*, 30(1): 184–212.
Davis, J. (2002). *Stories of change: narrative and social movements*. Albany: State University of New York Press.
Douglas, M. (2005). *Purity and danger: An analysis of concepts of pollution and taboo*. London: Routledge.

Duncan, P. (2013). Russia, the West and the 2007–2008 electoral cycle: did the Kremlin really fear a 'coloured revolution'? *Europe-Asia Studies*, 65(1): 1–25.

Ekho Moskvy, 2009. Tekst ekspertizy psikhologa Guzevoi. Ekho Moskvy [online], 11 September. Available from http://echo.msk.ru/blog/echomsk/619278-echo/ [accessed 15 March 2017].

Elias, N. (1978). *What is sociology?* New York: Columbia University Press.

Epshtein, A., Budraitskis, I. and Penzin, A. (2012). *10 let 'antiekstremistskoi' kampanii s Rossii*. Moscow: Svobodnoe Marksistskoe izdatel'stvo.

Epshtein, A. and Vasil'ev, O. (2011). *Politsiia myslei. Vlast', eksperty i bor'ba s ekstremizmom v sovremennoi Rossii*. Moskva: Gileia.

Gal, S. and Kligman, G. (2000). *The politics of gender after socialism: A comparative-historical essay*. Princeton: Princeton University Press.

Gaufman, E. (2017). *Security threats and public perception: digital Russia and the Ukraine crisis*. London: Palgrave Macmillan.

Gerring, J. (2006). *Case study research principles and practices*. Cambridge, UK: Cambridge University Press.

Goffman, E. (1986). *Frame analysis: An essay on the organization of experience*. Boston: Northeastern University Press.

Goffman, E. (1977). The arrangement between the sexes. *Theory and Society*, 4(3): 301–331.

Hemment, J. (2015). *Youth politics in Putin's Russia: producing patriots and entrepreneurs*. Bloomington: Indiana University Press.

Horvath, R. (2013). *Putin's 'preventive counter-revolution': post-Soviet authoritarianism and the spectre of Velvet Revolution*. London: Routledge.

Iuristy za konstitutsionnye prava i svobody. (2006). *Etnicheski izbiratel'nyi podkhod v deistviiakh militsii v Moskovskom metro*. Moscow: Novaia iustitsiia.

Johnson, P. (2011). Homosexuality, freedom of assembly and the margin of appreciation doctrine of the European Court of Human Rights: Alekseyev v Russia. *Human Rights Law Review*, 11(3): 578–593.

Kokorev, D., Marchenkov, A., Iashin, I., Shmakotin, M., Kozlov, A., Bol'shakov, I. and Kozlovskii, O. (2006). Politizirovannaia rossiiskaia molodezh': Pokhmel'e komsomolom – beremennost' grazhdanskim obshchestvom. *Neprikosnovennyi zapas*, 1: 89–126.

Kon, I. (2013). Lackmustest. Homophobie und Demokratie in Russland. *Osteuropa* 63(10): 49–67.

Kravchenko, M. (2013). *Inappropriate enforcement of anti-extremist legislation in Russia in 2012*. [Online.] Moscow: Sova Center for Information and Analysis. Available from www.sova-center.ru/en/misuse/reports-analyses/2013/06/d27382/ [accessed 10 March 2016].

Kravchenko, M. (2014). *Inappropriate enforcement of anti-extremist legislation in Russia in 2013*. [Online.] Moscow: Sova Center for Information and Analysis. Available from www.sova-center.ru/en/misuse/reports-analyses/2014/06/d29660/ [accessed 10 March 2016].

Kravchenko, M. (2015). *Inappropriate enforcement of anti-extremist legislation in Russia in 2014*. [Online.] Moscow: Sova Center for Information and Analysis. Available from www.sova-center.ru/en/misuse/reports-analyses/2015/06/d32083/ [accessed 10 March 2016].

Kuzio, T. (2005). Russian policy toward Ukraine during elections. *Demokratizatsiya*, 13(4): 491–517.

Lassila, J. (2014). *The quest for an ideal youth in Putin's Russia II. The search for distinctive conformism in the political communication of Nashi, 2005–2009*. Second Edition. Stuttgart: Ibidem.

Lerner, J. and Zbenovich, C. (2013). Adapting the therapeutic discourse to post-Soviet media culture: The case of Modnyi Prigovor. *Slavic Review*, 72(4): 828–849.

Lyon, D. (2009). Surveillance, power, and everyday life. In: C. Avgerou, R. Mansell, D. Quah and R. Silverstone (eds), *The Oxford handbook of information and communication technologies*. Oxford: Oxford University Press, 449–472.

Mann, M. (2004). *The dark side of democracy: explaining ethnic cleansing*. Cambridge, UK: Cambridge University Press.

McRobbie, A. (2009). *The aftermath of feminism: gender, culture and social change*. London: Sage.

Meylakhs, P. (2009). Drugs and symbolic pollution: The work of cultural logic in the Russian press. *Cultural Sociology*, 3(3): 377–395.

Monahan, T. (2010). *Surveillance in the time of insecurity. Critical issues in crime and society*. New Brunswick: Rutgers University Press.

Morozov, V. (2015). *Russia's postcolonial identity: A subaltern empire in a Eurocentric world*. Basingstoke: Palgrave Macmillan.

Mosse, G. (2010). *The image of man: The creation of modern masculinity*. New York: Oxford University Press.

Omel'chenko, E. (2006). Nachalo molodezhnoi ery ili smert' molodezhnoi kul'tury? *Zhurnal issledovanii sotsial'noi politiki*, 4(2): 151–183.

Oppenhäuser, H. (2011). Das Extremismus-konzept und die Produktion von politischer Normalität. In: Forum für kritische Rechtsextremismusforschung (ed.), Ordnung. Macht. Extremismus. Effekte und Alternativen des Extremismus-Modells. Wiesbaden: VS Verlag, 35–58.

Pitts, C. and Ovsyannikova, A. (2015). Russia's new treason statute, anti-NGO and other repressive laws. *Houston Journal of International Law*, 37: 83–138.

Pohl, R. (2004). *Feindbild frau: männliche sexualität, gewalt und die abwehr des weiblichen*. Hannover: Offizin.

POLIT.RU. (2012). Chego khoteli i chego dobilis' Pussy Riot: diskussiia v tsentre molodezhnykh issledovanii NIU VShE 30 March. Available from http://polit.ru/article/2012/04/16/discussion/ [accessed 17 March 2017].

Polletta, F. and Ho, K. (2006). Frames and their consequences. In: R. Goodin and C. Tilly (eds), *The Oxford handbook of contextual political analysis*. Oxford: Oxford University Press, 187–209.

Riabov, O. and Riabova, T. (2014). The remasculinization of Russia? Gender, rationalism, and the legitimation of power under Vladimir Putin. *Problems of Post-Communism*, 61(2): 23–35.

Richter, J. (2009). The Ministry of Civil Society? The public chambers in the regions. *Problems of Post-Communism*, 56(6): 7–20.

Rossoliński-Liebe, G. (2014). *Stepan Bandera: The life and afterlife of a Ukrainian nationalist*. Stuttgart: Ibidem.

Schwenck, A. (2015a). 'Antifaschistische Bewegung' als Selbstbezeichnung [no pagination]. Available from www.dekoder.org/de/gnose/antifaschistische-bewegung-als-selbstbezeichnung [accessed 31 March 2017].

Schwenck, A. (2015b). *Regierungsfinanzierte Jugendbewegungen in Russland* [no pagination]. Available from www.dekoder.org/de/gnose/regierungsfinanzierte-jugendorganisationen [accessed 31 March 2017].

Sedgwick, M. (2010). The concept of radicalization as a source of confusion. *Terrorism and Political Violence*, 22(4): 479–494.

Somers, M. and Gibson, G. (1993). *Reclaiming the epistemological 'other': narrative and the social constitution of identity.* Ann Arbor: University of Michigan.

Spektorowski, A. (2003). The New Right: ethno-regionalism, ethno-pluralism and the emergence of a neo-fascist 'Third Way'. *Journal of Political Ideologies,* 8(1): 111–130.

Sperling, V. (2015). *Sex, politics, and Putin: political legitimacy in Russia.* New York: Oxford University Press.

Stepanenko, V. (2005). 'Oranzhevaia Revoliutsiia' – priroda sobytii i osobennosti natsional'noi grazhdanskoi aktivnosti. *Vestnik Obshchestvennogo Mneniia: Dannye. Analiz. Diskussii,* 6(80): 25–38.

Surovtsev, I. (2013). Gosudarstvennaia Molodezhnaia Politika: Sostoianie i problemy. In: N. Litvinov and I. Surovtsev (eds), *Molodezh i ekstremizm v Rossii: informatsionno-analiticheskii sbornik,* Vol. I. Voronezh: Ekspertnyi tesentr Natsional'noi bezopasnosti, 3–8.

Swidler, A. (1995). Cultural power and social movements. In: W. Gamson, H. Johnston and B. Klandermans (eds), *Social movements and culture.* Minneapolis: University of Minnesota Press, 25–40.

Temkina, A. and Zdravomyslova, E. (2014). Gender's crooked path: feminism confronts Russian patriarchy. *Current Sociology,* 62(2): 253–270.

Urban, M. (1994). The politics of identity in Russia's postcommunist transition: The nation against itself. *Slavic Review,* 53(3): 733–765.

Voitsekhovskii, A. and Tkachenko, G. (2008). 'Oranzhevye' – nositeli neofashizma. In: Iu. Kozlov (ed.), *Banderizatsiia Ukrainy – glavnaia ugroza dlia Rossii.* Moscow: Iauza press, 371–382.

Wiebe, K. (2013). Verordnete verunsicherung. Homosexualitat, jugendschutz, jugend-literatur. *Osteuropa,* 63(10): 99–105.

Legislation

Government of the Russian Federation. (2014). *Rasporiazhenie: Osnovy gosudarstvennoi molodezhnoi politiki v Rossiiskoi Federatsii na period do 2025 goda.* Moscow: Gosudarstvennoe uchrezhdenie – izdatel'stvo 'Iuridicheskaia literatura'. Available from http://government.ru/media/files/ceFXleNUqOU.pdf [accessed 30 March 2017].

Government of the Russian Federation. (2006). *Rasporiazhenie: Strategiia gosudarstvennoi molodezhnoi politiki v Rossiiskoi Federatsii.* Moscow: Gosudarstvennoe uchrezhdenie – izdatel'stvo 'Iuridicheskaia literatura'. Available from http://government.ru/docs/all/58323/ [accessed 18 March 2016].

Chapter 13

Biocultural metrics and the moral policing of young people's politics in contemporary India

Pramod K. Nayar

This chapter examines the policing of young people's politics in India. Moral policing, a direct result of what I term cultural surveillance, targets the behaviour of young people which is a priori deemed dissenting, rebellious and threatening to the social order rather than as experimental or imbued with its own set of risks and uncertainties. It argues that the embodied – as in the corporeal – making, remaking and restricting of contemporary public space in India is characterised by the rise of biocultu/ral metrics.

Moral policing is a form of governance that involves normative but often non-official and extra-legal injunctions laid down, mostly, by non-state actors and vigilante outfits, driven by specific attitudes towards young people, the social order and the nation. The state in India, for instance, does its own mode of moral policing in the form of censorship of films and television programmes and laws governing alcohol sales in pubs and bars. However, since the mid-1990s, there have been more instances of such 'action' against 'erring' young people by vigilante groups such as the Hindu Yuva Vahini, the Rama Sene, Dukhtaran-e-Millat (restricted to Kashmir) and, more recently, cow-protection groups. Other organisations, such as the Shiv Sena, and political parties have sought to impose moral codes of conduct: asking for bans on Valentine's Day celebrations, protesting against air-kissing as a mode of greeting, intimate scenes in books and films, music bands, sex education in schools, young men and women partying together and even, at times, the clothing (shorts or skirts) of sports stars on the field (in 2015, a women's football match had to be called off in Kolkata because the local *maulvis* objected to the shorts the players were expected to wear).

These acts of moral policing have been, it should be noted, indulged in by *both* Muslim and Hindu vigilante groups, and regional political parties, in the past. The vigilante groups often align themselves with political parties although they seem to function with a fair degree of autonomy, sometimes to the embarrassment of the party itself. Such groups often identify 'Westernisation' as the bogeyman that 'Indian culture', both Hindu and Muslim, has to battle. This battle is carried out in the realm of culture, and cultural surveillance is the direct actionable programme of the moral police.

Surveillance 'involves assorted forms of monitoring, typically for the ultimate purpose of intervening in the world' (Haggerty and Samatas 2010: 2). It is a preliminary to social sorting (Lyon 2001, 2003). In the case of cultural surveillance, the intervention is in the form of legislating upon and thus organising 'youth behaviours', styles and attitudes through the threat of vigilante action and the making of normative codes in these realms. Those deemed to fit the idea(l) of a 'good Indian citizen' are separated from those who do not: defined as girls in Western outfits, women who drink at pubs, youth who indulge in public displays of affection or celebrate events like Valentine's Day.

Cultural surveillance is a policing of young people's politics, especially if we consider the emergence of new forms of social conduct, styles and entertainment amongst college-going or young professionals as 'politics'. This is not politics in the sense of seeking administrative or financial power, but politics as an expression, usually embodied, of a set of aspirations and attitudes of an entire generation. Commentators have noted that cultural fables of transformation – embodied, say, in make-over TV, fashion advice, self-help books – in consumer culture are disguised instances of the new managerialism and therefore constitute a politics (Rimke 2000; Ahuvia 2002; Hancock and Tyler 2004; Bratich 2007; Hearn 2008; Ouellette and Hay 2008). Cultural well-being in civic society then becomes the focus of intellectual debates and even active intervention, such as those by the state or vigilante groups. It is therefore in the domain of culture and cultural practices, where the younger citizens seek to forge their identity, community and sense of belonging, united by shared cultural fables and codes (whether of fashion, behaviour or taste in music), that moral policing and cultural surveillance make their intervention.

Public spaces and the rise of biocultural metrics

Cultural surveillance is inevitably a public phenomenon, targeting young bodies in public spaces. Almost all actions of vigilante groups and moral police have been in public spaces. Cinema houses, pubs and restaurants, college campuses, parks and recreational spaces have been the focal points for activities of cultural surveillance. Why public spaces, and why young bodies? My argument has two elements, one concerning the nature of public space and one concerning the forms of surveilling young bodies in that space.

Public spaces are spaces where an *assembled* and *embodied* politics is played out, where power asserts itself upon certain forms of embodiment and where such power is resisted and contested (Butler 2015). The young bodies themselves may be read in terms of a collective embodiment because, as recent commentators adapting a new materialist approach have suggested, instead of focusing on individual human bodies we need to examine them 'within material assemblages of bodies, things, ideas and social institutions', to focus on 'the micropolitics of constraining forces on bodies, and the capacities these assemblages produce in bodies' (Alldred and Fox 2017: 4). I propose that the embodied making, remaking and restricting of contemporary public space in India is characterised by the rise of biocultural metrics.

While biometrics measure irises, gait, posture, facial expressions and skin patterns, amongst others, biocultural metrics is the evaluation of youth biologies and their cultural appurtenances and manifestations. It is the deployment of yardsticks of 'appropriate' behaviour of these bodies in public. Biocultural metrics determines the limits of agency of individual young men and women to choose their bodily presentation in everyday life and public spaces. Within the assemblages of bodies, ideas and institutions, biocultural metrics would involve the drafting of codes of dressing, which define how a bodily form may be clothed for public consumption. 'Metrics' suggests itself as a shorthand for the measurement and indexicality of 'Indianness' with which vigilante groups operate to curb 'offensive youth behaviour'.

The public space of the restaurant, the recreation zone, the mall and the college campus are sites where a transformation in cultural values is most visible. From looks to language, these are spaces where the youth *express* themselves, mostly in terms of their bodies. Emerging cultural trends and the becoming-visible of hitherto hidden cultural practices attract attention precisely because these occur in the public domain. For instance, with the decriminalisation of homosexuality in 2009, Gay Pride parades made their first appearance in the public domain in India (Ray 2017). Commentators (Singh 2016) note that legalisation does not guarantee social acceptance. Thus, the moral policing of homosexuals in the public arena, leading to social stigmatisation and even stray incidents of violence, continued.

I suggest that the cultural surveillance of public spaces for 'dissident' cultural codes amongst youth is a battle for the very nature of *publicness* in a democracy such as India. Right of access, right of way and the right of/to assembly in the case of public spaces is linked to the very idea of the *demos* of democracy, especially because the 'public' has always been a space of discrimination, restriction and inhibition for certain sections of society. As commentators have noted (Phadke *et al.* 2011), the public space in India is discriminatory towards the woman flâneur, for instance. Full, free and safe public access remains a distant dream, a condition that erodes the foundations of a full, free and safe *demos* itself.

The public is the space where specific models of 'the Indian citizen' are being displayed, contested and negotiated, and this is primarily the province of youth cultures. Valentine's Day, an increasingly popular 'day' for youthful partying and romances, has become the target of attacks and criticism by self-professed guardians of 'Indian' culture. The Shiv Sena's secretary, Vinayak Raut, declared in 2011: 'We will not allow such days to harm Indian culture' (Deshpande 2011). Partying and public displays of affection, clothing and flash mobs, now frequently seen in India, are cultural trends where different practices are *practised* in public. The larger point is that the young people with these newer cultural practices are asserting their right to such practices in public: an idea that immediately draws the moral police.

Heterogeneity of cultural practices that we now see in such public spaces directs the attention to shifting cultural trends, and the fact that 'heterogeneity of the earth's population is an irreversible condition of social and political life itself' (Butler 2015: 112). Self-fashioning in the public space as cosmopolitan citizens of India attracts opprobrium in the form of moral prohibitions. In other words, the diversity of cultural practices, including new, emergent and radical ones, *symbolises the diversity that makes up the demos*. Cultural surveillance is the attempt to develop norms of regulating this diversity. It seeks to ensure a homogeneity founded on a presumption of uniformity of cultural practices, especially across generations. The conflict over public articulations of different cultural practices is therefore a form of non-official/non-state governance.

Embodied youth politics may be read as attempts at the fashioning of a recursive public. A recursive public is:

> a public that is vitally concerned with the material and practical maintenance and modification of the technical, legal, practical and conceptual means of its own existence as a public; it is a collective independent of other forms of constituted power and is capable of speaking to existing forms of power through the production of actually existing alternatives.
>
> (Kelty 2008: 3)

Flash mobs and newer cultural practices are therefore political practices that signal the publicness of the public, where the boundaries of the public and its freedoms are being guarded against vigilantism and as sites of expressive, enunciative action. When the 'youth' object to vigilantism and cultural surveillance, they are in effect surveilling the very conditions in which they can function as a public and as full citizens of India. Youth cultures such as the above evidently employ their bodies as part of their public self-representation. Thus, embodied practices are political acts that enunciate agency within the assemblage of institutions, affect, ideas and established cultural practices. They are instances of *biocultural* metrics because, after all, the young people's bodies are at once biology and culture. When young bodies seek freedom within these assemblages – for example, flash mobs or the 'Kiss of Love' campaigns (2014, 2017) where young men and women across various Indian cities kissed in public as part of a non-violent protest against the moral policing of youth – then they generate a micropolitics of a kind that redefines the public space itself.[1]

Concomitantly, cultural surveillance also focuses on the *body* of the youth: its comportment, fashion, stylisation and behaviour in *public* spaces, seeking to enforce a different biocultural metrics upon them. In contemporary India, the assemblages within which young bodies are enmeshed include a resurgent right wing (both Islamic and Hindu), a rapidly expanding consumer citizenship for all and more 'secure' spaces such as malls in which to hang out, amongst others. An increasing awareness of fashion trends and newer varieties of recreational pursuits – from bowling alleys to go-karting – also marks the young Indian. It is within this assemblage that we also see revanchist and constraining forces at work.

The bodies of young people are examined for their *manner of publicness* in cultural surveillance. By this, I mean the manner in which bodies fit into the assemblages. That is, many young people use their bodies to redefine the nature of public space itself. The response to this in the form of cultural surveillance takes recourse to biocultural metrics. As I write this, a news report states how a village in the Mathura district of the northern Indian state of Uttar Pradesh, through its all-male governing body (the Panchayat), has decreed that girls will be fined INR 2,100 for using their cell phones in public places (*The Hindu* 2017). This was not, it was clarified, a ban on their owning or using cell phones within the (monitored) spaces of their home, but a restriction on public usage: in order, the village elder said, to 'prevent girls from eloping with boys'. This is an instance of biocultural metrics because it calibrates and computes telephone usage, location of usage and possible moral, social and cultural effects of the usage. Cultural surveillance here is the restriction on the use of technology ostensibly directed at controlling youthful libido but effectively, also, the behaviour of the village's young girls in public spaces.

Most of the recent incidences of cultural surveillance have targeted girls' clothing, partying, socialising with boys. The vigilante groups target young couples as well. A gender bias is clearly evident when the girl is seen as the repository of cultural values, and therefore has to be protected from 'bad' influences. This insistence on the bodies of youth as signifying a 'national culture' is made possible by a process termed 'signification spiral'.

Signification spiral, moral crisis and civilising offensives

Actions and behaviours – such as girls speaking on cell phones in public – are bestowed with an increasing level of signification, a process that scholars of moral panics term 'signification spirals'. It is 'a way of publicly signifying issues and problems which is intrinsically escalating, i.e. "it increases the perceived potential threat of an issue through the way it becomes signified"' (Hall and Jefferson 1977, cited in Thompson 1998: 16). Charles Krinsky, in his gloss on the concept, argues that 'less dangerous activities become identified as symptoms or precursors of superficially similar but more destructive behaviours' (2013: 6).

In 2009, pubs in Mangalore were attacked because young men and women were drinking together (*Times of India* 2009a). Pramod Mutalik, whose Rama Sene led the attack, declared that they, the attackers, were 'custodians of Indian culture' (*Times of India* 2009b). The implication might be read in terms of a signification spiral: young people drinking together was 'unIndian'. From the act of consumption to the alleged violation of an 'Indian' norm of social and gendered behaviour, we see the meaning of the act escalate into a different idiom. The act of consumption is seen as symbolic of a larger crisis of Indianness itself. In the case of Islamic outfits, especially in places like Kashmir, vigilante groups issued injunctions on how the Muslim women of the region should dress

(Bukhari 2001). Consenting adults were also picked up from hotel rooms by the police in stray incidents of moral policing (Devnath 2015).

Instances such as the above vigilante actions are to be read within a context wherein 'the legitimacy of romantic love as a basis for marriage is itself in question' (Vanita 2005: 182). Thus, examples and spectacles of youthful romances in public spaces generate the debate whether 'romantic love undermines the social fabric or sustains it' (182). When young couples are harassed by vigilante groups (for instance, Rajeevan 2017), the subject in question is the legitimacy of romance. A signification spiral emerges when young romance is interpreted as marking the collapse of the (supposedly homogenous, self-evident) Indian social fabric itself, as the vigilante groups' rhetoric suggests.

Once a signification spiral is in place and assumes a self-perpetuating nature, the focus shifts to the emerging moral crisis. The use of cell phones, as the most recent case cited above suggests, enables romances to flourish and facilitates elopement. This is the perceived moral crisis: a rupture in the traditional modes of heterosexual partnerships (arranged marriages).[2] The premier institution of higher education, Aligarh Muslim University, prohibited students of the Women's College from accessing the university's main library in 2014. The reason for this ban, as given by the vice chancellor, was: the presence of girls in the library would attract more boys leading to a 'discipline issue' (Agha 2014). In 2015, Mahesh Sharma, a government minister, stated that: 'Girls wanting a night out may be all right elsewhere, but it is not part of Indian culture' (*Times of India* 2015). In all three cases, the behaviour of women, and signs of their empowerment – whether in the form of cell phones, access to educational resources or a 'night out' – are amplified in symbolic significance so that they represent the moral crisis faced by Indian culture.

The romance of an individual couple, even across religious or ethnic divides, is resignified as a moral crisis. Minor events 'gain strong visibility thanks to the newsworthiness of the established news theme, making them appear – to journalists as well as the public – to be new occurrences of the ongoing emergency' (Maneri 2103: 176). Youthful romances – the stuff of Bollywood – are recast as demonic, disruptive, disrespectful (of cultural values) and therefore newsworthy as a cultural emergency.

The authorities' fear of 'disciplinary issues' that may arise when boys and girls meet – in libraries, no less! – is an instantiation of a moral panic and a cultural crisis. The emphasis on discipline signals the perception of a threat.

Frank Furedi, writing about contemporary fear cultures, notes:

> Unlike conventional forms of moral panic, which were directed against clearly recognisable folk devils, twenty-first century anxieties are frequently focused on phenomena that are less tangible and less familiar but whose impact is seen to have catastrophic consequences. The current catastrophic imagination is also drawn towards a growing range of potential 'conditions' that threaten human extinction. Outwardly at least, fear appeals about the

dangers of global terrorism, superbugs, or global warming appear to be very different to the traditional targets of moral panics (e.g., juvenile delinquents, drug addicts, single mothers).

(2011: 96)

Furedi's point might be expanded to speak of the *cultural catastrophes* that authorities such as vigilante groups and the state authorities invoke through the signification spiral. Therefore, it could be suggested that moral crises are reinterpretations of youth cultural patterns as cultural catastrophes, which in turn entail a greater surveillance of 'youth behaviour'. The assumption that library spaces would only be used by young men and women for romance, rather than for studies, renders the young as gullible, wayward and frivolous. The politics of segregation, restricted access and outright prohibition around young people assumes that their behaviour is already a cultural threat, and that their personal behaviour – whether romance or flâneurie – is political in and of itself.

'Discipline', as the byword for discriminatory practices and moral policing, such as the ban on women in public libraries, may then be seen as 'civilising offensives'. In her work on moral panics, Amanda Rohloff has argued:

With moral panics, there need not be an actual weakening, only a perceived weakening. This could include the perception that governmental regulations, and the enforcement of those regulations, are failing to control a particular perceived problem; or, conversely, that individuals are failing to regulate their own behaviour and therefore there is a need for a stronger external force [...] to 'control' these 'uncontrollable' deviants.

(2011: 74–75)

Such regulation takes the form of 'civilising offensives', wherein the deviants and the uncontrollable have to be brought back into the civilisational field. Rohloff writes: while '"civilizing offensives" can be compared with projects of moral regulation (and, by extension, episodes of moral panic), they may involve within them a *fusion* of civilizing and decivilizing trends' (77, emphasis in original).

In contemporary India, civilising offensives are not always legal measures, but often a mix of statist institutional regimes and extra-legal forms of regulation. Thus, we can consider the Aligarh Muslim University's prohibition of women readers in its library or the village governing body's ban on public cell phone use by its young girls as civilising offensives wherein the young women are being trained to accept and conform to norms of cultural behaviour.

The argument that these so-called civilising or 're-educating' projects are also decivilising may be read off from the context where in the age of equal opportunities and women's empowerment campaigns there appears a simultaneous revanchism in circumscribing the behaviour and attitudes of the young women. Take, for instance, the campaign by both Hindu and Muslim vigilante groups against inter-community socialising, romances and marriages. A string of attacks on such romances has occurred across India since the 2000s (Verma 2014). More recently, the Hindu Yuva Vahini has been accused of killing a Muslim man, suspected of having helped his (Muslim) neighbour elope with a Hindu girl (*The Wire* 2017). In some cases, the state even stepped in to stop this 'love jihad' – a term used to speak of Muslim boys allegedly luring Hindu women into romance and marriage – from spreading (*Deccan Herald* 2009). The attempt to curb such liaisons, one commentator argues, 'amounts to complete control over the Hindu girls' movements, who they meet, where they go and reveals an anxiety over women going out much more and making independent choices, irrespective of caste or religion' (Gupta 2014). Thus, the civilising offensives are in the long term detrimental to women's rights and thus constitute a reversal for the numerous campaigns for gender justice, and may be read as decivilising.

In this process of civilising/decivilising, the girls are suddenly transformed into 'innocent victims' of 'cunning' Muslim young men in an astonishingly simplistic narrative. An organisation, Hindu Behen Beti Bachao Sangharsh Samiti, set up helpline numbers for 'distressed' parents of Hindu girls who had been so lured away (Ali 2014). The use of melodramatic constructions of distressed parents, gullible girls and devilish young men ensures that a cultural practice amongst young men and women – socialising in mixed company in public spaces – is viewed through a moral lens. Joanne Westwood has noted that in order to draw attention to child trafficking as a moral issue, the UK press has been using 'melodramatic tactics to arouse public indignation and anger at the exploitation of innocent victims' (2015: 90). We could, following Westwood, argue that the civilising offensive is also directed at fuelling public indignation through this process of constructing female victims.

Such rhetoric effectively rules out any agentic role for the girls themselves, and positions the male crusaders from the vigilante outfits as rescuers and protectors. By invoking threats towards 'innocent girls' in this fashion, and by putting in place 'disciplinary' measures (on clothing, social interaction, leisure activities, even technologies) in the guise of both saving the girls and Indian culture, the civilising offensive effectively shuts out the recreational, social and public spaces to the girls.

Conclusion

Moral policing is cultural surveillance targeted at young people. It re-reads young people's actions, assembly, bodily comportment as geared towards difference, towards cultural othering and, therefore, as threats to the (mono)cultural order or as contaminants of pure cultural forms.

Moral policing, of which I have offered instances from contemporary India, may be read as civilising offensives, founded on purely speculative biocultural metrics, against a supposed cultural threat embodied in the assembly and public presence of young bodies. However, these offensives carry within them a homogenising regime that seeks to circumscribe young people's action and activities, and therefore may be read as forms of governance masked as cultural 'projects'.

The contemporary cultural surveillance ethos suggests that youth seeking to assert themselves in public spaces, and even to redefine those spaces, are to be monitored. In this process, claims and rights to public space are at stake, and the contest between independent, agentic youth behaviour and statist-authoritarian, vigilante forces is a battle, finally, for the ways in which India will define its demos, and its citizens. I have argued elsewhere (Nayar 2016) that student protests in public universities which spill over into the spaces of the city may be read as instances of public pedagogy. Through such protests, theories and ideas, ideals and ideologies, debates about continuing discrimination, identity, patriotism, freedoms move from the classroom to the public space of debate: the *polis*. Extending this argument, I propose now that youth behaviour embodies the ideals of a democratic polis and demos. How free and public is the polis and the demos is visible in the freedom and agency accorded the youth: and this is the politics of contemporary India.

The circumscription of young people's behaviours in contemporary India is a set of attempts by vigilante groups of various religious and ethnic persuasions to develop specific models of Indian citizenship. These models are, in most cases, a monocultural response to the forces of globalisation, hybridisation and the contingent politics of the postmodern. The guarding of cultural boundaries, in the form of injunctions against dress codes, entertainment spaces, romances, is a surveillance system put in place with the assumption that the young people's experiments are in and of themselves dissenting and fractious in consequence.

Notes

1 One needs to examine young people's: employment of their bodies, sexualities and attitudes in public/offline spaces with their employment in online spaces. Commentators have noted that, very often, girls have used the 'safe' environs of the digital world to position themselves as hypersexualised, which makes them 'a part of the hegemonic process of sexualization that ultimately leads to their marginalization' (Thiel-Stern 2014: 172).

2 The 'dangers' of romantic love that is projected as a moral crisis for Indian society may be read in terms of contexts as specific as the taboo on sex education in schools. Hegemonic femininities and masculinities are constructed and these are deemed inviolable. See the work of Chakraborty (2010).

References

Agha, E. (2014). 'Girls in AMU library will attract boys – VC'. *Times of India*, 11 November. Available from http://timesofindia.indiatimes.com/city/agra/Girls-in-AMU-library-will-attract-boys-VC/articleshow/45101345.cms [accessed 6 May 2017].

Ahuvia, A. (2002). Individualism/collectivism and cultures of happiness: A theoretical conjecture on the relationship between consumption, culture and subjective well-being at the national level. *Journal of Happiness Studies*, 3: 23–36.

Ali, M. (2014). 'Hindutva groups to continue with campaign against "love jihad"'. *The Hindu.* 18 September. Available from www.thehindu.com/news/national/hindutva-groups-to-continue-with-campaign-against-love-jihad/article6421500.ece [accessed 6 May 2017].

Alldred, P. and Fox, N. (2017). Young bodies, power and resistance: A new materialist perspective. *Journal of Youth Studies*, 20(9): 1–15.

Bratich, J. (2007). Programming reality: control societies, new subjects and the powers of transformation. In: D. Heller (ed.), *Makeover television: realities remodelled.* London: IB Tauris, 6–22.

Bukhari, S. (2001). 'Kashmiri women bear the brunt again'. *The Hindu.* 14 September. Available from www.thehindu.com/2001/09/14/stories/0214000k.htm [accessed 4 May 2017].

Butler, J. (2015). *Notes towards a performative theory of assembly.* Cambridge, MA: Harvard University Press.

Chakraborty, K. (2010). Unmarried Muslim youth and sex education in the bustees of Kolkata. *South Asian History and Culture*, 1(2): 268–281.

Deccan Herald. (2009). 'Karnataka to take steps to counter "love jihad" movement'. 22 October. Available from www.deccanherald.com/content/31814/karnataka-take-steps-counter-love.html [accessed 6 May 2017].

Deshpande, A. (2011). 'Shiv Sena to keep an eye on Valentine's Day celebrations'. *Daily News and Analysis.* 14 February. Available from www.dnaindia.com/mumbai/report-shiv-sena-to-keep-an-eye-on-valentine-s-day-celebrations-1507526 [accessed 4 May 2017].

Devnath, S. (2015). 'Mumbai: couples picked up from hotel rooms, charged with "public indecency"'. *Mid-Day.* 8 August. Available from www.mid-day.com/articles/mumbai-couples-picked-up-from-hotel-rooms-charged-with-public-indecency/16437186 [accessed 4 May 2017].

Furedi, F. (2011). The objectification of fear and the grammar of morality. In: S. Hier (ed.), *Moral panics and the politics of anxiety.* London; New York: Routledge, 90–103.

Gupta, C. (2014). 'Interview. Induleka Aravind'. *Business Standard.* 6 September. Available from www.business-standard.com/article/beyond-business/love-jihad-campaign-treats-women-as-if-they-are-foolish-charu-gupta-114090600699_1.html [accessed 6 May 2017].

Haggerty, K. and Samatas, M. (2010). Introduction: surveillance and democracy: An unsettled relationship. In: K. Haggerty and M. Samatas (eds), *Surveillance and democracy.* London; New York: Routledge, 1–16.

Hancock, P. and Tyler, M. (2004). 'MOT your life': critical management studies and the management of everyday life. *Human Relations*, 57(5): 619–645.

Hearn, A. (2008). Insecure: narratives and economies of the branded self in transformation television. *Continuum*, 22(4): 495–504.

Kelty, C. (2008). *Two bits: The cultural significance of free software.* Chicago; London: Chicago University Press.

Krinsky, C. (2013). Introduction: The moral panic concept. In: C. Krinsky (ed.), *The Ashgate research companion to moral panics.* Farnham: Ashgate, 1–16.

Lyon, D. (2001). *Surveillance society: monitoring everyday life.* Buckingham: Open University Press.

Lyon, D. (2003). *Surveillance after 9/11.* Cambridge: Polity.

Maneri, M. (2013). From media hypes to moral panics: theoretical and methodological tools. In: C. Critcher, J. Hughes, J. Petley and A. Rohloff (eds), *Moral panics in the contemporary world.* New York: Bloomsbury, 171–192.

Nayar, P. K. (2016). How UoH and JNU have taken us from public protest to public pedagogy. *The Wire.* 20 April. Available from https://thewire.in/26990/how-uoh-and-jnu-have-taken-us-from-public-protest-to-public-pedagogy/.

Ouellette, L. and Hay, J. (2008). Makeover television, governmentality and the good citizen. *Continuum*, 22(4): 471–484.

Phadke, S., Khan, S. and Ranade, S. (2011). *Why loiter: women and risk on Mumbai streets*. New Delhi: Penguin.

Rajeevan, R. (2017). 'Women's Day in Kerala: Shiv Sena men chase and threaten couples with canes'. *India Today*. 8 March 2017. Available from http://indiatoday.intoday.in/story/kerala-moral-policing-shiv-sena-womens-day-aditya-thackeray/1/900103.html [accessed 5 May 2017].

Ray, S. (2017). 'Mumbai Pride march: LGBT supporters share their hopes for the future'. *Hindustan Times*, 11 February. Available from www.hindustantimes.com/mumbai-news/mumbai-pride-march-lgbt-supporters-share-their-hopes-for-the-future/story-XbUxZfVSCZ4DSpOTAXHs2H.html [accessed 3 May 2017].

Rimke, H. (2000). Governing citizens through self-help literature. *Cultural Studies*, 14(1): 61–78.

Rohloff, A. (2011). Shifting the focus?: Moral panics as civilizing and decivilizing processes. In: S. Hier (ed.), *Moral panics and the politics of anxiety*. London; New York: Routledge, 71–86.

Singh, P. (2016). Between legal recognition and moral policing: mapping the queer subject in India. *Journal of Homosexuality*, 63(3): 416–425.

The Hindu. (2017). 'U.P. village to fine women using phone'. 4 May. Available from www.thehindu.com/todays-paper/tp-national/no-mobiles-diktat-for-women/article18380345.ece [accessed 28 May 2017].

The Wire. (2017). 'Hindutva's "love jihad" obsession leads to murder of Muslim man in U.P'. 3 May. Available from https://thewire.in/131678/hindutvas-love-jihad-obsession-leads-to-murder-of-muslim-man-in-up/#disqus_thread [accessed 6 May 2017].

Thiel-Stern, S. (2014). *From the dance hall to Facebook: teen girls, mass media, and moral panic in the United States, 1905–2010*. Amherst: University of Massachusetts Press.

Thompson, K. (1998). *Moral panics*. New York; London: Routledge.

Times of India. (2009a). 'Girls assaulted at Mangalore pub'. 26 January. Available from http://timesofindia.indiatimes.com/city/mangaluru/Girls-assaulted-at-Mangalore-pub/articleshow/4029791.cms [accessed 6 May 2017].

Times of India. (2009b). 'Rama Sene on pub attack'. 27 January. Available from http://timesofindia.indiatimes.com/city/mangaluru/Ram-Sena-on-pub-attack-Were-custodians-of-Indian-culture/articleshow/4033150.cms [accessed 6 May 2017].

Times of India. (2015). 'Night out for girls not in our culture, says Union culture minister Mahesh Sharma'. 19 September. Available from http://timesofindia.indiatimes.com/india/Night-out-for-girls-not-in-our-culture-says-Union-culture-minister-Mahesh-Sharma/articleshow/49019712.cms [accessed 6 May 2017].

Vanita, R. (2005). *Love rites: same-sex marriage in India and the West*. New Delhi: Penguin.

Verma, L. (2014). '"Love jihad": ABVP to form vigilante groups on UP campuses'. *Indian Express*. 21 September. Available from http://indianexpress.com/article/cities/kolkata/love-jihad-abvp-to-form-vigilante-groups-on-up-campuses/ [accessed 6 May 2017].

Westwood, J. (2015). Unearthing melodrama: moral panic theory and the enduring characterisation of child trafficking. In: V. Cree, G. Clapton and M. Smith (eds), *Revisiting moral panics*. Bristol: Policy Press, 83–92.

Chapter 14

Surveillance and the student

Government policing of young women's politics

Paromita Sen

No Warden as Protector, No Brother who is tough
My Feminist Comrades Are Enough!!
Pinjra Tod![1]

Though 'students' mobilisation has long been a prominent feature of movements for political reform around the world' (Weiss *et al.* 2012; see also Rhoads 2016), India's student population has faced significant challenges in recent years from governments aiming to reduce dissent and opposition. Paradoxically, in an 'age of surveillance', where governments worldwide use big data, CCTVs and other sophisticated technologies to control their citizens, the governance techniques and ways in which surveillance is used to govern young people's politics in India remain more traditional. These traditional tactics are particularly apparent in the government policies directed at the political actions of young women enrolled in institutions of higher learning. The reliance on these traditional methods is especially surprising in light of the creativity and innovation displayed by students mobilising and pushing back against government encroachment into their daily lives. In this chapter, I look at a women's movement involving students enrolled in higher education, #PinjraTod,[2] and their clashes with the government. Though the movement originated in response to questions of housing and accommodation for women, the scope of the Pinjra Tod movement has expanded to encompass collective student interests, and thus highlights the intertwined relationship between government surveillance and student mobilisation in India more generally.

Surveillance and state repression

While research on the political repression of social movements has often focused on the role of police or physical repression (see Davenport 2005; Sheptycki 2005; Fernandez 2008; della Porta and Reiter 2013), Starr *et al.* (2008: 252) argue convincingly that 'other forms of repression have been less visible and not well understood'. Starr *et al.* focus on the impact of surveillance on mobilisation and free association, finding that surveillance of social movements has a destructive influence on young people's access to resources, opportunities for progressive action, recruitment and political consciousness. Fernandez (2008) found that state repression of social movements through surveillance is successful whether it deploys covert or more overt means, for example violent forms of repression involving police action. Furthermore, fear of being surveilled also has a dampening effect on collectives and individuals. For instance, Davenport (2005) and Boykoff (2013) found that organisations shift towards defensive practices when members fear they are being surveilled. Other researchers found that surveillance can push individuals away from overt collective resistance (i.e. protests and rallies) towards forms of covert resistance (Johnston 2005), or in some cases, towards more violent forms of resistance (della Porta 1995). However, Heynen and Van Der Meulen (2016: 24) argue that 'systems of surveillance are deeply gendered' and call on us to further investigate the intersectional impacts of surveillance. For instance, Kovacs (2017) analyses the gendered nature of surveillance, arguing it disproportionately reduces women's mobility and access to freedom, thereby limiting their political activism.[3] According to Koskela (2002), CCTV surveillance erodes women's confidence in institutions of power, while Law and Bruckert (2016) argue that surveillance through social networks (the surveillance web at strip clubs) allows for even those without technological capacity to participate in surveillance tactics and extend forms of control onto marginal bodies.

This chapter extends these arguments to the issue of student mobilisation in India. I argue that the government uses surveillance rather than physical policing as a method to repress and control women's movements, like the Pinjra Tod movement. The reliance on surveillance allows the government to achieve the same impact as technologically sophisticated surveillance but with much lower administrative costs. I draw together arguments from research on state repression of social movements, surveillance and student mobilisation to present new empirical data on how state repression of student movements looks in the modern day.

Pinjra Tod and questions of university access

Indian universities typically encompass colleges within their administrative purview and, for the most part, are public institutions controlled by government rules. These colleges very often do not offer on-campus housing (referred to as 'hostels' in India, and known as 'college dormitories' in other countries) for their students. A majority of students must find their own accommodation while enrolled in college. Additionally, where hostels are offered to students, different rules apply for men and women who reside in these hostels, including but not limited to curfews, dress codes and personal technology policies.

Scholars over the years have argued that students are uniquely equipped to effectively mobilise and demand attention – 'they are comparatively free of career and family obligations that might raise the stakes for acting out' (Weiss *et al.*

2012: 5) – and the shared space of the university provides the physical means to overcome the collective action barrier more easily. From its origins in localised campus struggles for equitable gender access to hostel accommodation, Pinjra Tod has succeeded in building a significant social movement. Its initial struggles were restricted in scope, in the numbers of participants and in the limited nature of their victories. However, in September 2015, a small protest at a university in Delhi became the catalyst for a nationwide movement involving women students from across the country, all demanding equitable access to higher educational resources. The participants were women otherwise separated by geography, class, caste, religion and socio-economic background, but coming together in support of equal access to education, and bound by their shared sense of alienation from institutional educational spaces. We see in Pinjra Tod the successful genesis of a national movement, one which achieves small-scale incremental but steady victories while developing a practice of activism and an increasingly heightened awareness of the student body's latent political potential.

For instance, while the movement has primarily been framed as fighting for gender-neutral housing policies and equitable access to educational resources, it has also proved to be a successful catalyst extending beyond what might be strictly considered women's causes. It has become a key player in political activism intended to keep college campuses secular, resist the tide of oppressive nationalism currently sweeping India, shield minorities from caste and religious-based violence and protect educational funding from government attempts to privatise and corporatise higher education (Pinjra Tod (@pinjratod)). So far, it has done this while retaining its political autonomy and remaining independent from political affiliation with university bodies and organisations like Women's Development Cells and other women's NGOs. However, the movement has aroused hostility from other students and from administrators and government officials. For instance, Pinjra Tod members were physically attacked at a protest by members of the student wing of the ruling BJP party (Kausar 2017). Members have been threatened with rape by elected members of student government in response to their mobilisation on multiple occasions (interviews with Pinjra Tod members[4]).

Paradoxically, in responding to a movement that draws heavily on new technology and social media to mobilise and sustain action, the government has primarily relied on tactics that draw on family structures, conservative ideologies and cultural repertoires. My research suggests that the government uses traditional surveillance techniques such as curfews and dress codes managed by hostel supervisors. Indian governments have become increasingly aware in recent years of the advantages of doing this as they set out to manage student mobilisation. In the case of the Indian government's response to mobilisation by women students, this starts by framing such mobilisation as antithetical to the safety and interests of the women students.

Winter (2008: 6) defines framing as 'the process by which political leaders communicate about issues by emphasising certain features of an issue, downplaying others, and assembling those features into a coherent narrative with clear implications for policy action'. Important to note here is that 'Frames specify *how* to think about things, but they don't point to *why* it matters' (Ferree and Merrill 2000: 485). While much of the framing literature focuses on how activists have mobilised to challenge and resist governments and institutions, I find here that the government also uses framing to co-opt and recreate narratives about public mobilisation. In effect, governments have begun using the very tools created to fight against them as ways to strengthen their political bases and grip over the state. The Indian government, for instance, justifies withholding the right of women students to access space freely through a concern for their safety and an interest in ensuring they focus on their academic study. In this way, the right to education that students mobilised to demand is utilised by the state as justification for their imprisonment – that is, the frame proposed by Pinjra Tod to define itself is now co-opted by institutions of power against the movement itself.

While the effectiveness of these methods of repression is yet to be determined, the Pinjra Tod movement is just one of many forms of student action the Indian government is attempting to eradicate through attacks on their leaders and ordinary members. In later sections of this chapter, I offer a brief overview of the higher education context in which the Pinjra Tod movement and state action against it belongs. I then detail the various methods of repression used by the government against the women of the Pinjra Tod movement, treating them as emblematic of attacks on politically active young women. I conclude by considering the implications of government surveillance and repression for the future of young people's politics in India.

Methodology

In this chapter, I draw primarily on a series of semi-structured interviews with participants in the Pinjra Tod movement, conducted over a period of 18 months.[5] I also rely on the movement's own records, as its members have documented various measures adopted by the universities in their attempts to police young women. I also draw on a 2015 report compiled by the movement for the Delhi Commission for Women (DCW) which documents the history of the struggle for equitable access to education. Additionally, I use relevant secondary data culled from newspaper coverage. The movement is largely made up of students, many of whom are researchers and scholars at leading research institutions. Their reports and records reflect a high level of scholarly rigour and critical analysis of the situation. Finally, I draw on field notes and observations from a number of campaign events I attended over a period of two months in 2015 and another two months in 2017.

Hostel access – questions of power and privilege

There are close to 30 million students currently enrolled in over 700 universities and 35,500 affiliated colleges in India, of whom 46 per cent (or 13 million) are women (Choudaha 2013). Delhi University's various colleges have over 132,000 students enrolled in undergraduate and postgraduate courses (University of Delhi website). Many of Delhi University's students come from all over the country, seeking higher education and upward social mobility because of its reputation as one of the nation's premier universities. However, in 2015, only 9,000 spaces were provided at the university's hostels (for men and women). Furthermore, these are granted on the basis of academic ranking rather than need – so students who score better in end-of-school exams are offered hostel accommodation first. Universities do not therefore take into account a student's socio-economic class or individual needs when providing hostel access.

This system has immediate consequences for students. Access to on-campus accommodation is often cited as a determining factor for students and their families when deciding which college to enrol in. Several interviewees recall having to choose a lower-ranked college because it provided hostel accommodation, despite being offered admission to a higher-ranked college where they would be required to find alternate housing. This problem is particularly severe for women, as their parents tend to trust university housing to 'protect their daughters' as opposed to independent housing in the city. Interview participants[6] recall friends and cousins (all women) who were not permitted by their parents to study in Delhi at all because they did not secure hostel housing. The rules of hostel access create barriers for young people from the beginning, especially for those rooted in vulnerable communities. Aware of this dependence on university housing, the government often uses hostel access as leverage to control student behaviour and repress their opposition to university and government policies.

Preventive backlash

In the aftermath of the now infamous 2012 brutal rape and murder of a college student in Delhi (Sengupta 2012; Zitzewitz 2013; Lodhia 2015), several Indian universities chose to implement even more stringent, primarily gender-specific, rules regarding hostel residency. The spectre of the brutal rape allowed university administrations to roll back basic civil liberties – a move justified under the guise of protecting their female wards. While this particular action was not in direct response to the Pinjra Tod movement, it rolled back the victories eked out by the movement to ensure equitable hostel access in Indian higher education nonetheless.

The newly updated accommodation regulations included, but were not limited to, strict curfews and dress codes, mandatory local guardians (registered residents of the city whose approval was required to participate in any non-hostel-sanctioned activities outside college grounds), removal of locks from room doors[7] and installation of CCTVs within the hostels, including near bathrooms. While these governance measures uniformly targeted women, they had a different impact on different groups of women. For instance, the impact of the new rules regarding local guardians is felt more severely by poorer students, students from outside the city and students from more conservative families. Students come to Delhi in pursuit of higher education from every state in the union. Those from poorer backgrounds or rural communities are less likely to have family in the city, as are those from the states furthest away from Delhi. Many students must rely on very distant family friends or relatives to serve as their local guardians, thus creating yet another hurdle for those from traditional families, and those with fewer resources. As one student recalled, her parents had to spend over a month visiting various relatives and friends of family, asking them to take responsibility for her. Students often found themselves with local guardians they had never met prior to joining college. These guardians tended to be wary of giving their 'wards' permission to extend curfew, as they worried the students might hurt their family's 'reputation' – a concern often raised by university administrators when meeting local guardians. As such, the practice of requiring guardians served to further marginalise already vulnerable populations, who already experienced limited access to the city due to barriers imposed by their gender, caste, class or ethnicity.

The 2015 Pinjra Tod report provides details on the oppressive rules at Indraprastha College for Women, which included strictly enforced dress codes in the cafeteria, ostensibly to protect 'the modesty' of the women students in the presence of male staff working in the cafeteria (Delhi Commission of Women Report 2015). Hindu College similarly required a dress code ostensibly to ensure the 'safety of women students' and to 'maintain decorum' (Sen 2016). Additionally, restrictive curfews were imposed on women residents at university hostels that required women to be back in the hostels by 6.30 p.m. in the evening in winters, and 7 p.m. in the summers.

Earlier protest actions by women students at university hostels had succeeded in having the curfew time delayed till 9 p.m., and student residents were optimistic that they could improve on those advances. However, in the aftermath of the 2012 rape and the accompanying administrative overhaul of hostel administration, the curfew for women across all dormitories in the university settled on 6.30 p.m. curfews. The Union Minister of Women and Child Welfare went so far as to say that curfews are a 'necessary safeguard' that provide protection from teenage 'hormonal outbursts', and therefore needed to be enacted to secure the virtue and purity of 'Indian youth' (Choudhury 2017).[8]

The consequences of such a curfew are manifold. With curfew set for mid-evening, women students are unable to attend most cultural events, social events and/or even go to the libraries and laboratories. Furthermore, private landlords who rent rooms to women students follow the government's policies and surveil their tenants, use CCTVs and impose rules about dress codes and no-visitor policies. The consequences of this are that university campuses essentially become all-male spaces in the evening. Even if some women are not bound by curfews, the absence of so many women from the streets makes many women who are on the streets feel unsafe and vulnerable to harassment, as women who occupy public space after dark are considered promiscuous (Kovacs 2017).

As an administrative measure, the curfew creates divisions between women students, whereby those who disobey the curfew or have lenient guardians, are labelled as 'promiscuous' and therefore responsible for any violence they might experience (Law and Bruckert 2016). This creates further divisions between students and working women outside the college space entirely. By prohibiting women in colleges from using public spaces, the state makes the streets more unsafe for women who have to use public space late at night for a variety of reasons, such as those who work late and/or rely on public transport to bring them home. Through these governance measures, the state succeeds in labelling some women as worthy of protection (those in higher education, those with families who will impose curfews, those who can afford to rent a private room from a landlord) while the others are left to fend for themselves on streets that are systematically being made more unsafe.

Policing tactics

What happens to women when they organise and resist? Amongst the punitive measures available to governments is moral policing, which draws on traditional family structures and social structures to define *appropriate behaviour* in contrast to *transgressive behaviour*. Moral policing is a form of governance that draws its power from societal structures and customs, and the student embeddedness in social networks and families. Behaviour is monitored and punished through a reference on *moral* codes as opposed to legal ones. For instance, women are punished for missing curfew by hostel administrators (often referred to as wardens in the Indian context), not through fines, demerits or other administrative sanctions, but with threats of calling parents with (often false) stories about sexual promiscuity. The involvement of parents in the disciplining process reflects a twin strategy of both shaming and infantilising as forms of governance. Shaming women who miss curfews, for example, both shames the woman for her adult choices and dismisses her as not being adult enough. A primary tool in the warden's arsenal is to call a student's parents and convey (exaggerated, if not blatantly untrue) reports of transgressive behaviour, primarily commentary on the student's sexual behaviour. As punishment for missing a curfew by 15 minutes, one interviewee recalled her father being phoned and told she had slept with all the men at a neighbouring boys' college. In another instance, the 400 hostel residents of Patiala Girls College unanimously signed a petition demanding revision of their 6.30 p.m. curfew; in response, the administration informed all parents of this petition. Though the students were all legally adults, their 'transgressive' behaviour was deemed to be the parents' responsibility. Many parents sided with hostel administrators and supported the suspension of their daughters, their denied access to the outside world post-curfew and some even endorsed additional punishment in some instances. Only students whose parents agreed they could extend the curfew were allowed an extension to stay out beyond the designated curfew time.

Such practices have significant political consequences for the development of young people's politics in India, specifically amongst women. Participation in democratic practices like a protest can result in consequences for students; official complaints to families that more often rely on false reports are designed to malign the student and, in some cases, result in the loss of their hostel accommodation for the following year. They are practices that also thwart the process by which some young women attempt to gain independence from their families, as any attempt to 'be independent' is reported to their families. In this way, the university is literally surveilling students and reporting to families. In turn, parents tend to interpret these official reports from universities as evidence that their children aren't taking their education seriously enough. As one student recalls, her roommate was withdrawn from college by her family due to her participation in protests about lack of amenities in her hostel, because her parents viewed her participation in politics as evidence of her lack of dedication to her education.[9]

While moral regulation is effective, other popular strategies to govern young women's politics rely on the law. On admission to a hostel, students and their parents must sign a residency contract. Much of it is standard, stating, for instance, that the student will not partake in drugs, alcohol or bring weapons into the dormitory. However, such contracts also require that the student not criticise university policy in any form whatsoever, or partake in political activities that can be construed as critical of the university or its benefactors. Considering that most universities in India are public institutions, this is often used to silence democratic opposition to governments and institutions.

These contracts are also often changed arbitrarily, with no explanation provided, so students are frequently unaware of which rights they have signed away. One interviewee's attempt to stage a student production of Eve Ensler's *Vagina Monologues* saw her 'written up' because the material was said to be 'obscene'.[10] The fact that the student signed the contract was cited by the university and used to coerce her into dropping her plans to organise the production. While she agreed, she claims she could not recall having signed away the right to participate in dramatic performances. Pinjra Tod has found that many students and their families are unfamiliar with this contract system due to their social background, are not informed of their rights when signing these documents and do not fully understand their content or appreciate their consequences.

University administrators have also found other ways of denying hostel accommodation to students deemed to be transgressive. For instance, St. Stephen's College, one of India's premier colleges, now requires mandatory interviews for admission every year. Such a requirement works to intimidate and discourage student participation in protests as such action is considered to be grounds for removal from the hostel. For instance, the annual re-interview process was recently used to take away hostel access from students involved in protesting against the college administration's attempt to become autonomous from Delhi University. Autonomy for the college from the university would mean that

St. Stephen's College would not be bound by the rules concerning tuition and affirmative action that govern centralised Indian universities. This would therefore result in fee hikes and an end to the reservation system that ensures students from marginalised communities are provided places in colleges. Students committed to protecting higher education from what they perceived as privatisation (an attempt to capitalise on the reputation of the college to demand higher fees than mandated by the government) were punished by the revocation of their hostel privileges for the year (Pinjra Tod (@pinjratod) 2017a).

This approach to controlling and monitoring young people isn't all that hidden as an aspect of Indian politics. For instance, the St. Stephen's College principal claimed that 'curbing women's movement helps them improve academically and anyone opposed to the curfew is not serious about their studies' (Joshi 2013) – effectively framing dissent on college campuses as the work of students not serious about their study, and thus deserving of expulsion. This approach to criminalising students results in students being arrested for organising speaker forums, as well as being charged with sedition for protesting against a 1,100 per cent fee hike in one year (Pinjra Tod (@pinjratod) 2017b). What must be noted here is that the criminalising of student mobilisation is not restricted to universities in rural areas, or to universities promoting a conservative ideology. This narrative of 'unpatriotic' students has pervaded urban institutions and the best educational establishments, including the universities where political leaders in the past have begun their own political careers.

Hostels and beyond

This is not to say that young people and politics in India are not flourishing. Pinjra Tod continues to fight against hostel curfews across Delhi and the country, and celebrates every victory it achieves. However, the movement has also broadened its scope, since its inception. While still fighting various forms of gendered institutional harassment, it has now become allied with other movements that are emerging to fight other forms of institutional harassment. For instance, Pinjra Tod has been active in protesting against the violence meted out by educational institutions to Dalit and Muslim students. A 'lower caste' student described her feelings about the suicide of a fellow Dalit PhD student through the Pinjra Tod platform, and exhorted her fellow students to:

> Build a platform for solidarity for only the oppressed will understand one another. The upper caste may sympathise but they will not understand us. We need to build a bond among the Dalit Bhaujan Adivasi community.[11]

The anniversary of the brutal rape and murder of a 17-year-old Dalit student in rural Bihar, India, also became a rallying point and site from which collective criticism was voiced to authorities and a society unconcerned with the violence perpetuated against marginalised citizens (Pinjra Tod (@pinjratod) 2017c).

Some Pinjra Tod members argue that the case of a missing Muslim student, Najeeb, who went missing the day after he had a confrontation about political opinions with members of a Hindutva right-wing political party, is an example of the religious violence that is becoming more common in India, where Muslims are being punished for ostensibly transgressing religious boundaries.[12] Therefore, Pinjra Tod found common ground with other students protesting institutional insensitivity to Najeeb's disappearance, a shared alienation from institutional spaces.

Pinjra Tod also joined forces with and documented its participation in a wide range of allied causes, for example in solidarity with working women on International Working Women's Day, against campus sexual violence, against military impunity in Kashmir and the North East and against a trend of increasing campus violence disproportionately targeting minorities.[13] Although these protests have moved beyond the hostel curfew issue, members of the Pinjra Tod collective are finding and creating new resistances aimed at reimagining the university in the face of the increasing radicalisation of Indian politics more generally. The repression of women's rights across India's universities has in many cases created opportunities for the creation of shared narratives and strengthened alliances.

This shared resistance does, however, increase the risk of young women being targeted for attack, including the threat of violence. The increasing encroachment of India's right-wing Hindutva government into university spaces, largely carried out by their sponsorship of the student body party ABVP, has divided university student debates about the public sphere and freedom, with the right wing locked in opposition to all other political student groups in Indian universities. Pinjra Tod itself has actively campaigned against ABVP and criticised its oppressive politics. In response, ABVP members have physically threatened protesting students, and beaten up those they oppose. The mobilisation of political action by Pinjra Tod has also been met with direct threats by ABVP party cadres, involving threats of rape. A participant involved in putting up posters for the movement in its initial days had to file a police report against an ABVP cadre who threatened her with rape in response to her activism. Yet Pinjra Tod continues to resist, even in the face of such violence.

Conclusion

The Pinjra Tod movement continues to resist oppressive government policies. The movement's reliance on decentralised organisation is one of the few key elements that allows it to avoid the deleterious effects of surveillance on its organisational strength and cohesion. India's universities are fast becoming a microcosm of the larger battles fought out in India's democracy. Indian universities have typically been accessible to even those from the poorest backgrounds, as colleges were designed to be democratically affordable to all granted admission. Restricting access to universities by welcoming only populations the government views as 'worthy' of higher education is reflective of larger governmental efforts to reshape the Indian population more generally. Marginalisation of vulnerable populations has become integral to the way the government interacts with its citizens, and a reliance on traditional narratives to do so allows the government to label attempts to resist these policies as 'pro-Western', and thus 'unpatriotic'. The rise of the surveillance state has far-reaching consequences for the nature of Indian democracy in the twenty-first century. As Jasbir Puar argues, surveillance 'has certainly not become more democratic: who receives discipline and punishment, who is deemed worthy of pleasure and intimacy remains distributed in deeply uneven manners' (Puar, cited in Kovacs 2017). Kovacs (2017) observes that surveillance is pre-emptive – surveillance 'can also shape what you will do in the future. That is because surveillance can incentivise certain kinds of behaviour, and discourage others'. The continued surveillance of students will eventually impact on the capacity of students to mobilise themselves, thus undermining one of the key sources of resistance to the power of India's elites. Decentralised opposition that evades surveillance, is likely therefore to be one of the few successful tactics of resistance available to young people in India.

Notes

1 Pinjra Tod campaign poster.
2 Pinjra Tod literally means 'Break the Cage' in Hindi.
3 Sewell *et al.* (2016) find that overt surveillance – e.g. aggressive policing – has a disproportionate effect on men's psychological well-being and mental health more generally.
4 Interviews conducted October 2015.
5 Anonymised interview transcripts available upon request.
6 Interviews conducted in October 2015, May 2016, June 2016, and March 2017.
7 Perceived by many to be shrouded in homophobic diktats, whereby allowing women to lock their doors will result in women becoming lesbians (interviews with hostel residents).
8 When it comes to evolving policy, politicians and bureaucrats responsible for women and child welfare across the country follow a similar logic. For instance, The Karnataka Legislative Assembly's Women and Child Welfare Committee Chairperson, N. A. Harris, recently recommended that IT companies in the IT Hub of India, Bangalore, should hire more men, so as to ensure women didn't have to work night shifts. This was included in a report aimed at investigating solutions to make Bangalore safer for working women (Ananya 2017).
9 The interview participant believes the parents are currently searching for a groom for her ex-roommate while the woman herself is currently enrolled in a secretarial course.
10 Ironically, this occurred the same year that the university principal organised a campus-wide event where Eve Ensler, author of the *Vagina Monologues*, was the keynote speaker.
11 Lama (2017). @pinjratod Instagram post, 16 March.
12 Najeeb Ahmed, a MSc student at Jawaharlal Nehru University, went missing on 15 October 2016. The night before, he had been involved in a brawl with ABVP

members. While possibly unrelated, the JNU administration has been accused of apathy and bias in handling the issue, and of obscuring the brawl he had been involved in. As of April 2016, there have no major breaks in the case, but rumours abound. Many of these rumours have been negative – such as an alleged Google search for ISIS, his dependence on prescription medicines to treat mental health disorders, and reported presence in Muslim universities across the north (BBC 2016; Kidwai *et al.* 2016; Mallick 2017).

13 The Pinjra Tod Instagram and Facebook pages document their participation, involvement and often their messaging to those beyond the movement.

References

Ananya, I. (2017). 'How do we make night shifts safer for working women? Bangalore MLA has a hilarious suggestion'. *The Ladies Finger* [online], 28 March. Available from http://theladiesfinger.com/women-night-shifts/ [accessed 3 April 2017].

BBC News. (2016). 'The mystery of the missing Indian student'. BBC India [online], 26 October. Available from www.bbc.com/news/world-asia-india-37759469 [accessed 25 March 2017].

Boykoff, J. (2013). *The suppression of dissent: how the state and mass media squelch US American social movements.* New York: Routledge.

Choudaha, R. (2013). *Statistics on Indian higher education 2012–2013.* University Grants Commission of India. Published online at DrEducation: Global Higher Education Research and Consulting (www.dreducation.com/2013/08/data-statistics-india-student-college.html) [accessed 24 March 2017].

Choudhury, S. (2017). 'Maneka Gandhi: curfew in hostels to protect you from "hormonal outbursts"'. *NDTV* [online], 7 March. Available from www.ndtv.com/india-news/why-hostels-need-early-curfew-hormonal-outbursts-says-maneka-gandhi-1666742) [accessed 17 April 2017].

Davenport, C. (2005). Introduction. Repression and mobilization: insights from political science and sociology. In: C. Davenport, H. Johnston and C. Mueller (eds), *Repression and mobilization.* Minneapolis, MN: University of Minnesota Press, pp. vii–xii.

della Porta, D. (1995). *Social movements, political violence, and the state: A comparative analysis.* Cambridge: Cambridge University Press.

della Porta, D. and Reiter, H. (2013). Policing protest. In: D. Snow, D. della Porta, B. Klandermans and D. McAdam (eds), *The Wiley-Blackwell encyclopaedia of social and political movements.* Hoboken, NJ: Wiley-Blackwell.

Fernandez, L. (2008). *Policing dissent. Social control and the anti-globalization movement.* New Brunswick, NJ: Rutgers University Press.

Ferree, M. and Merrill, D. (2000). Hot movements, cold cognition: thinking about social movements in gendered frames. *Contemporary Sociology,* 29(3): 454–462.

Heynen, R. and Van Der Meulen, E. (2016). Gendered visions: reimagining surveillance studies. In: E. Van Der Meulen and R. Heynen (eds), *Expanding the gaze: gender and the politics of surveillance.* Toronto: University of Toronto Press, 3–32.

Johnston, H. (2005). Talking the walk: speech acts and resistance in authoritarian regimes. In: C. Davenport, H. Johnston and C. Mueller (eds), *Repression and mobilization.* Minneapolis, MN: University of Minnesota Press, 108–137.

Joshi, M. (2013). 'St. Stephen's hostel curfew stays, students miffed at "autocracy"'. *Hindustan Times: New Delhi,* 3 April.

Kausar, H. (2017). 'Ramjas College protest highlights: clashes between ABVP, DU students, cops crack down'. *Hindustan Times: New Delhi*, 6 March.

Kidwai, A., Rao M., Hariharan V., Nair, J., Dutta P., Kumar, A., *et al.* (2016). Continued absence of Najeeb Ahmad—letter to the editor. 22 October. *Economic and Political Weekly*, 51(43): [no pagination]. Available from www.epw.in/journal/2016/43/web-exclusives/continued-absence-najeeb-ahmad%E2%80%93letter-editor.html?0=ip_login_no_cache%3De283907302788930701a0bfd35911077 [accessed 15 February 2017].

Koskela, H. (2002). Video surveillance, gender, and the safety of public urban space: 'Peeping Tom' goes high tech? *Urban Geography*, 23(3): 257–278.

Kovacs, A. (2017). *Reading surveillance through a gendered lens: some theory.* Internet Democracy Project [online]. Available from https://genderingsurveillance.internetdemocracy.in/theory/ [accessed 17 April 2017].

Lama, S. (@pinjratod). (2017). 'I've been thinking for quite some time now…' Instagram. 16 March.

Law, T. and Bruckert, C. (2016). The surveillance web: surveillance, risk and resistance in Ontario strip clubs. In: E. Van Der Meulen and R. Heynen (eds), *Expanding the gaze: gender and the politics of surveillance.* Toronto: University of Toronto Press, 240–263.

Lodhia, S. (2015). From 'living corpse' to India's daughter: exploring the social, political and legal landscape of the 2012 Delhi gang rape. *Women's Studies International Forum*, 50(May–June): 89–101.

Mallick, A. (2017). 'Kidnapped, missing, or still in JNU: where is Najeeb Ahmed?' *The Quint: India.* 20 March.

Pinjra Tod. (@pinjratod). Instagram account. Available from www.instagram.com/pinjratod/

Pinjra Tod. (2015). *Break the hostel locks #pinjratod: end discriminatory restrictions on women.* Report submitted to Delhi Commission for Women.

Pinjra Tod (@pinjratod). (2017a). '360 resident students out of the 400 students of St. Stephens College…' Instagram, 9 May. Available from www.instagram.com/p/BT3q9w9FvuR/?taken-by=pinjratod [accessed 27 May 2017].

Pinjra Tod (@pinjratod). (2017b). 'Update from Panjab University, and the curious case of sedition'. Instagram, 12 April. Available from www.instagram.com/p/BSxtecjgdZt/?taken-by=pinjratod [accessed 27 May 2017].

Pinjra Tod (@pinjratod). (2017c). 'Exactly one year back, on this day, came the devastating news…' Instagram, 29 March. Available from www.instagram.com/p/BSOmhvaASH-/?taken-by=pinjratod [accessed 27 May 2017].

Rhoads, R. (2016). Student activism, diversity, and the struggle for a just society. *Journal of Diversity in Higher Education*, 9(3): 189.

Sen, J. (2016). 'Safety or moral policing? The unapologetic gender bias of Indian hostels'. *The Wire: Education.* 27 April.

Sengupta, A. (2012). 'Girl raped in bus, tortured with rod – Delhi victim on ventilator after attack'. *Telegraph: National.* 17 December.

Sewell, A., Jefferson, K. and Lee, H. (2016). Living under surveillance: gender, psychological distress, and stop-question-and-frisk policing in New York City. *Social Science & Medicine*, 159: 1–13.

Sheptycki, J. (2005). Policing protest when politics go global. Comparing public order policing in Canada and Bolivia. *Policing and Society*, 15: 327–352.

Starr, A., Fernandez, L., Amster, R., Wood, L. and Caro, M. (2008). The impacts of state surveillance on political assembly and association: A socio-legal analysis. *Qualitative Sociology* 31(3): 251–270.

University of Delhi. *About DU*. Available from www.du.ac.in/du/index.php?page=about-du-2 [accessed 25 March 2017].

Weiss, M., Aspinall, E. and Thompson, M. (2012). Introduction: understanding student activism in Asia. In: M. Weiss and E. Aspinall (eds), *Student activism in Asia: between protest and powerlessness*. Minneapolis, MN: University of Minnesota Press.

Winter, N. (2008). *Dangerous frames: how ideas about race and gender shape public opinion*. Chicago, IL: University of Chicago Press.

Zitzewitz, K. (2013). 'A timeline of events in the Delhi gang-rape case'. *The Feminist Wire* [online], 2 February. Available from www.thefeministwire.com/2013/02/a-timeline-of-events-in-the-delhi-gang-rape-case/ [accessed 28 November 2017].

Chapter 15

Electoral engineering and surveillance

British young people and politics

Matt Henn, Ben Oldfield and Judith Bessant

Public concern has accompanied evidence of a deepening disconnect between young people and formal politics in advanced democracies (Grasso 2016). In Britain, this divide is particularly visible in a decline in young people's electoral participation (Henn and Foard 2014; Henn and Oldfield 2016). Some commentators and academics mistakenly represent young people as apathetic – or anti-political (for overview, see Hay 2007; Stoker 2006). However, at least since the 2008 recession (Grasso and Giugni 2016), many young people have been actively engaged in 'alternative' political action of various kinds often seen by governments and other power elites as problematic (Pickard and Bessant 2017).

In this context, many governments considered their options to encourage young people's electoral participation. Their declared aim was to increase the turnout of perennial abstainers (i.e. 'the young', and the least politically interested) (Singh 2014). Various electoral reforms were proposed, including extending the voting period, increasing the range of voting methods beyond physical attendance at the ballot booth and introducing compulsory voting.

Henn and Oldfield (2016) reported on responses by some young people to these 'reforms'. In this chapter, we consider the value of these proposals to stimulate young people's electoral participation and whether they represent a new form of governance and surveillance. We begin by examining young people's disengagement from formal democratic politics and consider whether this is symptomatic of entrenched apathy and disillusionment, or indicative of a generation redirecting its attention towards alternative forms of political practice (Grasso 2013). We identify some of the proposals intended to enhance their electoral participation after which we assess young people's views of Britain's democratic processes.

Our research suggests that decisions by many 18-year-olds *not* to vote is not the product of apathy or of self-exclusion. Rather, many report they are anxious that there are few opportunities available for them to meaningfully participate in formal politics. They report that those elected to public office have failed to represent their interests and the issues they consider important. Evidence from data collected from a national survey of British 18-year-olds and focus groups (Henn and Oldfield 2016) supports the proposition that young people's disengagement from formal politics is due in part to their interest in new styles and forms of politics, as well as a degree of alienation from conventional politics (Henn *et al.* 2002).

In the final section of this chapter, we consider whether the discussion and promotion of new forms of electoral technologies can be understood as a process designed to introduce new forms of governance and surveillance. We identify some theoretical considerations that suggest why this might prove to be a fruitful field for further research.

Young people's (dis)engagement?

There is no doubt the electoral turnout of young people in Britain has until the recent 2017 election been in decline since the 1990s. But the 2017 contest was special – young people bore the brunt of government austerity policies and were increasingly frustrated with a political class that had ignored their concerns; they had also been attracted by the opposition Labour party's radical manifesto as well as its participative campaigning style (Henn and Hart 2017). Previously, it had been estimated that significantly less than half (43 per cent) of registered 18–24-year-olds had voted during general elections held in 2001, 2005, 2010 or 2015 (Ipsos MORI 2015), well below participation rates recorded during the 1980s and 1990s, and considerably less than for the older generations. Evidence also indicates that young people in the UK and other liberal democracies are withdrawing from other formal methods of political participation, and this is 'directly, or indirectly, linked to the functioning of political institutions' (Hooghe and Marien 2014: 538).

What explains this apparent young citizen–state disconnect? One suggestion is that young citizens face certain 'start-up' problems that inhibit their engagement with formal democratic politics (Parry *et al.* 1992). Various contextual factors like increasing mobility, extended 'transitions to adulthood' problematise their 'life-courses', leaving little time for conventional politics (Phelps 2012). As young people enter later phases of adulthood, however, it is claimed they acquire the resources and space to engage with the formal political process (Norris 2003).

An alternative position argues there is a too-heavy reliance on conventional indicators to define engagement with politics such as electoral activity and support for, and membership of, mainstream political parties (Henn *et al.* 2002). This 'anti-apathy' approach (Phelps 2012) claims that a traditional understanding of politics fails to appreciate the diverse range of non-institutionalised political actions in which many young people participate (Bessant 2014).

Proponents of this 'anti-apathy' approach claim young people actively bypass conventional modes of political participation and are inclined to engage in non-institutionalised and extra-parliamentary political action (Pickard and Bessant 2017). They are less hierarchical options and considered by many young people to be more expressive, participative, flexible, meaningful and more effective (Farthing 2010; Furlong and Cartmel 2012). Importantly, the 'anti-apathy' approach challenges claims that young people are uninterested in politics. Rather, emphasis is given to evidence indicating that while many British young people have an aversion to the forms of politics practised by professional politicians at Westminster, they are interested in politics in its broadest sense (Henn and Foard 2014; Holmes and Manning 2013).

And while some claim actions like street-based protest, online activism of various kinds, digital satire, sit-ins, flash gigs, or resisting state violence are not politics but 'slacktivism', larrikinism, criminal riots or fold politics, such characterisations are problematic – not least because they rely on an overly narrow understanding of the political. Indeed, they overlook the political nature of the motives that move people to act. When Coleman (2014) asked the young people she was researching why they engaged in online activism, she found that most identified their interest in values like public accountability, the right to freedom of information, speech and privacy, or ideas like fairness and justice. Others pointed to political and moral emotions, like outrage about the effects of policies or indignation about the privileges claimed by organisations they oppose (Coleman 2014). Many historians have also examined the political character of protest like those by agricultural and industrial workers (Bohstedt 2010). They too highlight the need to identify the ideas that moved people – like 'moral emotions' and the 'moral economy' (Newburn 2015: 48). Younge also responded to claims that recent protests in Britain were not 'political', by arguing that this denies:

> ...the political nature of what took place, it simply chooses to only partially describe it ... When a group of people join forces to flout both law and social convention, they are acting politically.
>
> (Younge 2011; see also Newburn 2015: 987)

Young people who engage in non-institutionalised and extra-parliamentary political action protesting, for example, against neo-liberal austerity cuts (Giugni and Grasso 2015) or the erosion of democratic principles, are often expressing frustration in the face of the palpable unwillingness of their governments and other power elites to deal with 'the big issues' (Bessant *et al.* 2017) or outrage about 'situated injustice' (Pickard and Bessant 2017). And as Martin Luther King, leader of the 1960s American civil rights movement, argued, the way our reactions to injustice move us to take action:

> It is not enough for me to stand before you tonight and condemn riots. It would be morally irresponsible for me to do that without, at the same time, condemning the contingent, intolerable conditions that exist in our society. These conditions are the things that cause individuals to feel that they have no other alternative than to engage in violent rebellions to get attention.
>
> (King 1968)

Too often, the conventional framing of politics as rational action oriented to compromise or consensus ignores those occasions when the political takes on emotional, creative, disorderly, even disruptive characteristics.

This is precisely the social logic informing a narrow conceptualisation of politics that fails to do justice to the myriad ways young people now engage in. When thinking about young people and the public sphere (from which most are excluded), considering the asymmetric nature of generational power relations in the field of politics helps reveal why it is problematic that experts through their everyday habits of work reproduce and impose their definitions of politics in ways that dominate and exclude.

UK government responses to the citizen–state divide

In Britain, politicians and some researchers observed this trend by young people away from formal electoral politics with increasing concern. In the aftermath of the 2001 UK General Election, a House of Commons Select Committee stated:

> …we find it extraordinary that [the] collapse in electoral participation, put alongside other evidence on civic disengagement, has not been treated as a civic crisis demanding an appropriate response.
>
> (House of Commons Public Administration Select Committee 2001;
> see also, House of Commons Political and Constitutional Reform
> Committee 2014)

Others point to the drop in political participation signified by declining election turnouts and party membership while noting that levels of engagement in other forms like signing petitions is stable or increasing (Wilks-Heeg *et al.* 2012: 4). Meanwhile, Wagner (2011) points to 'long-term terminal decline' as politicians become less representative of their constituencies and as disillusioned citizens stop voting. Wagner (2011: 22) explains it in terms of the old formula 'welfare state and mass loyalty', of the 1950s and 1960s being displaced with new 'liberal society and citizen disaffection' evident by declining rates of participation in formal politics and increasing dissatisfaction with government performance. This 'democratic crisis' reflects 'the inability of … political modernity to develop an inclusive egalitarian – and just – way of dealing with economic matters' (Wagner 2011: 21).

We suggest it's a 'crisis' unlikely to disappear in spite of government bids to encourage voter turnout and political engagement with initiatives intended to increase political literacy and make voting easier. This is because such initiatives do not address the underlying substantive reasons why so many young people are disengaging from formal politics.

The introduction of statutory citizenship classes in schools in England in 2002 was prompted by concerns of the Labour government that young people lacked an inclination to participate in political processes (Tonge *et al.* 2012).

Other initiatives intended to 'modernise' electoral administration (like the Representation of the People Act 2000) approved experiments in local elections to make voting easier. As a consequence, pilot schemes were conducted across many local authorities between 2000 and 2007. They included remote electronic voting (e.g. by internet and text messaging), voting during different hours, on different days and over a number of days, and all-postal voting. Wattenberg (2008) contends that while not a cure-all, such reforms reduce generational disparities. Other research, however, suggests that the effect of these initiatives has been mixed (Boon *et al.* 2007). Meanwhile, the UK House of Commons Political and Constitutional Reform Committee (2014) and the Electoral Commission called for more work on electoral modernisation.

Finally, some called for a system of compulsory voting, but only for first-time voters. In this scenario, young people who have reached the minimum voting age and not formerly voted would be legally obliged to vote – be it a local or national election. This, it was argued, would 'kick-start a life-long habit of voting' (Birch *et al.* 2013: 21). Unsurprisingly, this proposal provoked considerable debate because it raises questions about democratic principles and civil rights (Fischer 2011).

A key argument supporting compulsory voting for young people is that it increases this generation's attendance at the ballot booths by approximately 30 per cent (Hill 2011). Furthermore, if young people are compelled to vote, that might facilitate a habit for positive democratic engagement (Franklin 2004). Evidence also suggests that compulsory voting may reduce generational disparities in electoral participation rates (Singh 2014); in particular, it may ensure that socio-economically disadvantaged groups are neither under-represented at the ballot booth nor under-estimated in the minds of politicians and policy-makers (Hooghe and Pelleriaux 1998).

It is therefore reasoned that, in delivering higher UK youth turnout rates, compulsory voting might oblige professional politicians to treat young people on a par with their older contemporaries (Birch *et al.* 2013). In essence, by helping to eliminate a generational *electoral* divide, it is deduced that this would have the effect of reducing generational social and economic *policy* inequalities. This might contribute to the creation of a virtuous circle, in that young people begin to recognise the latent power of their vote and of their influence over the political class – and this might help shape an enhanced positive predisposition to electoral participation. This is, however, a contested position (Lever 2008).

A major disadvantage with such a compulsory voting scheme is the way it singles out young people as different from the rest of the population, reinforcing age-based stereotypes of 'youth' as distinctly apathetic and requiring coercion as the only way they will perform their civic duties. A key assumption underpinning the proposal is that young people are inherently deficient, and a 'problem' that needs fixing. It is a framing that precludes the idea that young people's disaffection may have a legitimate basis – that the problem is rooted within the political process or democratic institutions.

It may be argued that compelling a young person to vote also reflects a limited concept of politics as mainstream electoral – and overlooks the concerns of those who feel no affinity with the parties on offer. In this context, compelling one group (young people) to vote has serious negative implications for the health of a democratic system. Forcing young people to vote may encourage a sense of resentment and even increase the susceptibility of some to populist parties and movements (Jensen and Spoon 2011).

Young people engaging with 'Politics'

It is important to establish why many young people now abstain from voting in British elections. We draw on qualitative and quantitative data to analyse interest in, and understanding of, politics and elections on the part of young Britons, as well as their attitudes to the political process and democratic institutions (Henn and Foard 2012). That work, carried out in 2011, produced two data sets, the first was derived from a national, representative online survey of 1,025 young people aged 18 at the time of the 2010 UK General Election. The timing is important because it provides insight into the views and reactions of this age group after their first opportunity as newly enfranchised citizens to gain experience of life under a new government – regardless of whether they opted to vote in the election.

Henn and Foard (2012) also convened 14 online focus groups totalling 86 participants who were deliberately selected as young people who had *not* voted at the 2010 General Election. The intention was to explore how non-voters, so often characterised as politically apathetic, conceptualised politics. The focus groups provided opportunities to examine how discourses of politics are constructed when young people express their views amongst peers.

Data from the online survey and online focus groups provide insights into why some young people avoid formal democratic politics and why many continue to abstain from voting in elections.

In the focus groups, participants were asked, '*If I say politics, what do you say?*' Responses indicate they tended to define 'politics' in a formal sense – as the politics of Westminster, centred on elections and main political parties – and they found this form of politics remote, unappealing and reliant on deception. A small selection of typical responses by participants includes:

- I feel its boring as it doesn't feel like it relates to me in any way shape or form
- a way of making you feel you have a say when you don't
- I always try and avoid politics because they all get me angry
- they break promises, they want our votes then cast us off like we didn't care in the first place. I think if I ever voted I would feel used for the vote.

Despite this characterisation of politics, the data reveal that young people study the world around them – and often feel strongly about contemporary issues, even if they don't necessarily consider those issues to be overtly 'political' in nature. This aligns with the anti-apathy approach outlined by Phelps (2012), suggesting that young people are interested in matters of civic and political consequence.

Focus group participants were asked, '*If you could change one thing about the world, what would it be, and why?*' This question was designed to understand the issues of primary concern to them, and their reasons for those concerns. As expected, the answers were varied, although participants demonstrated a solidaristic outlook and a broad commitment to social justice. For instance, a significant number of respondents prioritised the tackling of domestic and global inequalities. They also advocated introducing measures designed to reduce poverty ('lessen the gap between the richest and poorest people'; 'I would like equality throughout the world and would like to end poverty'). The emphasis on collective social justice was also revealed through their expressed desire to eliminate prejudice and discrimination: 'stop descrimination and racism as everyone is the same'; 'put a stop to bullying and unfair treatment'.

The focus groups also revealed clear signs of underlying anxiety about their immediate educational and employment prospects:

- Higher education and in particular, employment prospects at the end of it
- the cost of education, i think education should be free for everyone no matter were you frm
- employmnt for the youths really s an issue for me.

Prospects for achieving political influence

Although they appear to be interested in politics, young people in this study seemed to lack confidence in the political system, or what others refer to as 'external efficacy' (Westholme and Niemi 1986). Focus groups' data revealed they feel unable to translate this interest into participation in the formal political system – which they consider unresponsive to the actions of groups of citizens. They felt that there are relatively few meaningful opportunities available for themselves or other young people to access political power. The data indicated that young people feel a sense of fatalism with respect to this lack of perceived external efficacy, especially in relation to participation:

- the indivdual cant change it
- plenty of people have voiced thier opinion about all the different cuts some through protests ... it still hasnt changed circumstances so little old me would be useless
- [...] one cannot do it themselves, we need thousands upon thousands to make the slightest difference.

We recognise that since the focus group participants were intentionally selected as *non*-voters, we might expect them to hold such views on this matter; however, given that a considerable majority abstained at the 2010 General Election, these results provide important insights into why they feel demotivated to vote. Interestingly, there was a mixed response when they were asked to comment on *collective* political action, including protests. A considerable minority of young people highlighted the potential efficacy of such methods for advancing causes and facilitating positive political change – especially in contrast to the perceived value of *individual* political action:

- i think there is strength in numbers not individuals acting alone
- i believe if large enough numbers it could [be effective]
- if a big enough group of people stand up they might listen!

However, the predominant view from the focus groups was that even such collective actions had only limited potential for effecting political change. Their response to events such as the student demonstrations in 2011 and the Occupy movement was relatively resigned and stoical:

- protest all you want it wont change anything
- it doesnt work, 100's of protests can happen but nt one will work it takes more then people sitting protesting for anything to b done
- take a look at the education protests [...] we tried SO HARD [...] we lost.

In many respects, these pessimistic views seem to be underpinned by a strong sense across the focus groups that 'the political class' is an unresponsive elite:

- i dont feel we have a chance of making a difference to things we care about as the desisions have already been made for us
- it feels like the public have no say. the governemt already have everything planned out
- the only way the government will actually take notice of us will be if we acted in the same way the people in the arab uprisings did but what could be a good enough reason to act like that?

Confidence in the democratic process and the political class

While this group appears relatively fatalistic with respect to the scope and prospects for achieving political influence, they are broadly supportive of democratic ideals and processes. For instance, results from the online survey suggest a majority of young people in Britain are committed to the principle of voting (57 per cent), with a sizeable group expressing support for competitive elections (48 per cent).

The current generation of young people are, however, deeply averse to professional politicians. The survey indicates that 81 per cent of young people disapprove of the political class. Data from the focus groups reveal the source and extent of this ambivalence – and in some cases hostilities. Professional politicians are seen as remote and incapable of governing in an empathetic manner or with due regard for the interests and aspirations of young citizens:

- they seem to live in a totally different world to everyone else
- they dont think like the majority of the 'common person'
- i dont think they can really relate to students or youths … clearly theres a problem when students are protesting and youths are rioting.

This political class is considered to be cynical and self-serving, and lacking any serious commitment towards championing the concerns of young people:

- i think the politicians have lost the trust of our generation
- … the goverment just dose what it wants to do and dosent listen to what the public want they say they will make changes so you will vote for them and then dont back up there promises
- … once a aprty is voted into power their views often change, as seen this year when two largely apposing parties joined in a coalition.

Given this context, how should we understand government interventions to reduce the ongoing young citizen–state disconnect? One way of interpreting it is to ask whether we are witnessing the evolution of new styles of surveillance of young people's politics.

Theorising surveillance

Surveillance is now a key feature of contemporary industrialised societies; a practice carried out by states, private corporations and some universities and schools (Marsden 2014). According to Staples, surveillance is 'the act of keeping close watch on people' involving a 'watchful gaze' (1997: ix). It involves monitoring everyday life by collecting and processing personal data (Lyon 2001) whether identifiable or not for the purpose of influencing or managing those whose data have been garnered.

As Lyon observes, increasingly, it does not involve embodied persons watching each other (2001: 2). Surveillance incorporates everything from visual observation (e.g. CCTV), genetic testing, collecting biometric data and use of facial recognition systems, electronic monitoring (e.g. opinion polling, metadata analysis of emails and phone calls) and the collection of statistical data to observe trends in social conduct (like crime, unemployment) and so on. Most surveillance operates abstractly and at a distance: it can be characterised as remote and impersonal watching. In this way, it separates watching from witnessing. And increasingly, the watching, analysing and interpreting is automated and reliant on algorithms. In these ways, surveillance has become ubiquitous and normalised.

Traditionally, when political science engaged with political surveillance it tended to be framed as a defensible response to small, well-defined groups of dissidents (Cunningham 2004). This naivety became less feasible after 9/11 as states introduced antiterrorism legislation mandating mass surveillance (Scheuerman 2016). For instance, in the USA, the passage of the USAPATRIOT Act (Uniting and Strengthening America by Providing Appropriate Tools Required to Intercept and Obstruct Terrorism) removed barriers that allowed the government to intercept electronic mail transmissions and monitor web surfing, all without necessarily informing the individual of this surveillance.

It is now generally acknowledged that governments in Europe, Australia and the USA regularly subject citizens to surveillance of their electronic communication (Stoycheff 2016). Snowden's revelations also highlighted how the National Security Agency, through its PRISM project, engaged in mass online surveillance programmes in 2013, in the USA and Europe. Originally, PRISM was designed to monitor and collect the online communications of foreign nationals suspected of terrorist involvement. PRISM allows US intelligence agencies to direct backdoor access to the servers of companies like Google, Microsoft, Apple, Facebook, Yahoo, YouTube, Skype and others, introducing the potential for surveillance of Americans without warrants (Greenwald and MacAskill 2013).

Surveillance has also become a normal feature of workplaces, schools, homes and communities. The once private space of the home has been entered by home-based media as we create possibilities for fully integrated surveillance systems that track our consumer, and other, activities. In the home, keyboard strokes are harvested, media preferences are collected, individual interactions with information technologies are now stored, tracked and logged by platforms like Facebook and Google. Meanwhile, 'data miners' monitor the 'mood' of the Internet via real-time analysis of online conversations, blog-posts, Tweets, Facebook posts and other activity.

Foucault (1979) drew on Bentham's idea of the Panopticon as a metaphor for modern society that deploys surveillance to regulate the lives of citizens to promote social order. Bentham's panopticon referred to a design principle that allowed prison inmates to be observed while eliminating the visibility of the guards. Thus, prisoners, unsure of whether or not they were subject to guards' gaze, were always governed because they came to act at all times as if they were watched so as to avoid punishment. This became an internalised practice as they came to govern themselves.

Thus, for Foucault, the Panopticon entails a 'set of techniques and institutions for measuring, supervising and correcting the abnormal … which, even today, are disposed around the abnormal individual, to brand him and to alter him' (Foucault 1979: 199). In this way, a Foucauldian account of surveillance provides a useful framework for theorising the impact of perceived government surveillance on young people's political activity. It's a framework that identifies 'political deviants' (or those 'at risk' of becoming 'deviant') as primary targets of surveillance. This perspective also allows for a range of reactions to surveillance, like docility or political resistance.

Internet-based surveillance by governments conforms in many ways to a panoptic model as it provides highly effective disciplinary means requiring a few to watch the many while preventing the many from watching the few (Boyne 2000). Conversely, internet users can also watch their watcher. As Boyne argues, 'the machinery of surveillance is now always potentially in the service of the crowd as much as the executive' (2000: 301).

The ubiquity of surveillance presents other problems for liberal democracies. Describing various socio-technical relationships as surveillance says little about 'the dynamics of the control, resistance, emergence and development of surveillance practices' or whether 'surveillance is a symptom of, or a precursor to new kinds of social and political practices' (Ball and Haggerty 2005: 133).

All of this highlights the value of considering the contemporary context and need to develop research programmes that focus directly on the idea and practice of surveillance in relation to young people and their politics.

One way of thinking about such a project could involve recognising how prevailing discourses about the 'crisis of democracy' function implicitly to 'justify' 'the need' to maintain and increase surveillance of 'problem youth' because they express dissent from mainstream politics variously described (e.g. young Muslims, protesters and other 'potentially problematic' groups) as 'likely' to disrupt the political consensus.

'Youth politics' is also a concern for states that prioritise security. One obvious target of this governance of counterterrorism is invariantly referred to as the religious and political 'radicalisation' of young people (e.g. Muslim, Middle Eastern, Pakistani or African). It occurs in the public sphere where political opinion emphasises how Islam represents a threat to Western, secular, liberal democracy. In this context, surveillance has become critical to the post-9/11 'culture wars', focusing on religion, gender, race and nationalism.

The extent to which British police and counterterrorism agents have emulated US surveillance practices is not clear. We refer to projects that include the use of technologies like 'mosque crawlers' and undercover informants, 'rakers' (usually Muslims), deployed to identify 'suspicious' Muslim and Arab nationals as part of a 'human mapping programme' in co-operation with the CIA and drawing on Israeli surveillance techniques. In the US, such techniques were extended to protest actions like the Occupy Wall Street movement and Palestine solidarity rallies. As research reveals, ethnic minorities and those from low socio-economic backgrounds are more likely to withdraw from electoral participation which implies they are more likely than other groups of young people to be subject to state surveillance (Koch 2016).

With all this in mind, it may be pertinent to see the debates and research about the disengagement of young people from electoral politics and their interest in 'alternate' politics as an opportunity for reflexivity. The prevailing discourse of civic crisis has a long-standing connection with the self-portrait of the social sciences. Indeed, the social sciences as they developed through the twentieth century relied on the idea and promise they could help government manage recurrent social order problems and periods of crisis through empirical research into social, economic and political problems (Wagner 2001: 40–53).

One consequence of this is that the social sciences have become a critical part of the normalised surveillance technologies now employed by modern states. Recent versions of this are evident in the creation and deployment categories like 'youth at risk' and 'social capital' as researchers lament the alleged decline of social capital in late modern societies, and the ensuing apathy that is said to have replaced once close-knit communities and received forms of sociability and trust (e.g. Putnam 2004).

Conclusion

Care needs to be taken in accepting the tendency by some politicians, policy-makers and academics to present disenchantment with politics on the part of young people in terms of civic crisis and political apathy. Proposals to introduce age-specific compulsory voting may increase the numbers of young voters, but may also reinforce some young people's negative views about contemporary politics. If young people are to be motivated to engage with formal politics and the electoral process, then 'the political class' may benefit by looking beyond 'procedural-technical' programmes of the kind identified above. And, if young people are to be persuaded to participate in formal politics, the onus is on the political elites to think more creatively about how to make such democratic participation more meaningful and attractive to young people.

Mandating young people to vote to solve the problem of 'youth disengagement' from formal politics is problematic because it institutes forms of surveillance and governance that are significant for those interested in maintaining democratic values and practices like privacy and the ability to engage in deliberative practice without fear of coercion or the prospect of intimidation.

Everyday surveillance informed by an interest in governance of particular political actions presents particular problems for liberal democracies. 'Youth politics' and young people's withdrawal from electoral participation has become a major concern for states that prioritise security. In this context, we suggest it may be timely to see the prevailing debates about 'youth disengagement' from mainstream politics, their increased participation in 'alternate' politics and state responses to that as an opportunity for reflection. We say this because the 'civic crisis' discourses draw on a well-established relationship between the social sciences and modern states' various social reform projects informed by an interest in managing social order problems and crises through social scientific expertise.

References

Ball, K. and Haggerty, K. (2005). Editorial: doing surveillance studies. *Surveillance & Society*, 3(2): 129–138.

Bessant, J. (2014). *Democracy bytes: new media and new politics and generational change*. London: Palgrave Macmillan.

Bessant, J., Farthing, R. and Watts, R. (2017). *The precarious generation: A political economy of young people*. Abingdon, Oxon: Routledge.

Birch, S., Lodge, G. and Gottfried, G. (2013). *Divided democracy: political inequality in the UK and why it matters*. London: Institute of Public Policy Research.

Bohstedt, J. (2010). *The politics of provisions: food riots, moral economy, and market transition in England, c.1550–1850*. London: Ashgate.

Boon, M., Curtice, J. and Martin, S. (2007). *Local elections pilot schemes 2007: main research report*. London: The Electoral Commission.

Boyne, R. (2000). Post-panopticism. *Economy and Society*, 29: 285–307.

Coleman, G. (2014). *Hacker, hoaxer, whistleblower, spy: The many faces of Anonymous*. London: Verso.

Cunningham, D. (2004). *There's something happening here: The new left, the Klan, and FBI counterintelligence*. Berkeley: University of California Press.

Farthing, R. (2010). The politics of youthful antipolitics: representing the 'issue' of youth participation in politics. *Journal of Youth Studies*, 13(2): 181–195.

Fischer, C. (2011). Compulsory voting and inclusion: A response to Saunders. *Politics*, 31(1): 37–41.

Foucault, M. (1979). *Discipline and punish: The birth of the prison*. New York: Vintage.

Franklin, M. (2004). *Voter turnout and the dynamics of electoral competition in established democracies since 1945*. Cambridge: Cambridge University Press.

Furlong, A. and Cartmel, F. (2012). Social change and political engagement among young people: generation and the 2009/2010 British election survey. *Parliamentary Affairs*, 65(1): 13–28.

Giugni, M. and Grasso, M. (eds). (2015). *Austerity and protest: popular contention in times of economic crisis*. London: Routledge.

Grasso, M. (2013). The differential impact of education on young people's political activism: comparing Italy and the United Kingdom. *Comparative Sociology*, 12(1): 1–30.

Grasso, M. (2016). *Generations, political participation and social change in Western Europe*. London: Routledge.

Grasso, M. and Giugni, M. (2016). Protest participation and economic crisis: The conditioning role of political opportunities. *European Journal of Political Research*, 55(4): 663–680.

Greenwald, G. and MacAskill, E. (2013). 'NSA prism program taps in to user data of Apple, Google and others'. *Guardian*, 7 June [accessed 5 May 2017].

Hay, C. (2007). *Why we hate politics*. Cambridge: Polity.

Henn, M. and Foard, N. (2012). Young people, political participation and trust in Britain. *Parliamentary Affairs*, 65(1): 47–67.

Henn, M. and Foard, N. (2014). Social differentiation in young people's political participation: The impact of social and educational factors on youth political engagement in Britain. *Journal of Youth Studies*, 17(3): 360–380.

Henn, M. and Hart, J. (2017). [Forthcoming.] The generation election: youth electoral mobilisation at the 2017 General Election. In: D. Jackson, E. Thorsen and D. Lilleker (eds.), *UK election analysis 2017: media, voters and the campaign*. Poole: Centre for the Study of Journalism, Culture and Community at Bournemouth University.

Henn, M. and Oldfield, B. (2016). Cajoling or coercing: would electoral engineering resolve the young citizen–state disconnect? *Journal of Youth Studies*, 19(9): 1259–1280.

Henn, M., Weinstein, M. and Wring, D. (2002). A generation apart? Youth and political participation in Britain. *British Journal of Politics and International Relations*, 4(2): 167–192.

Hill, L. (2011). Increasing turnout using compulsory voting. *Politics*, 31(1): 27–36.

Holmes, M. and Manning, N. (2013). 'He's snooty 'im': exploring 'white working class' political disengagement. *Citizenship Studies*, 17(3–4): 479–490.

Hooghe, M. and Marien, S. (2014). How to reach Members of Parliament? Citizens and Members of Parliament on the effectiveness of political participation repertoires. *Parliamentary Affairs*, 67(3): 536–560.

Hooghe, M. and Pelleriaux, K. (1998). Compulsory voting in Belgium: An application of the Lijphart thesis. *Electoral Studies*, 17(4): 419–424.

House of Commons Political and Constitutional Reform Committee. 2014. *Voter Engagement in the UK, Fourth Report of Session 2014–15*. London: The Stationery Office Limited.

House of Commons Public Administration Select Committee. (2001). *Public Participation: Issues and Innovations: The Government's Response to the Committee's Sixth Report of Session 2000–01*. Available from www.publications.parliament.uk/pa/cm200102/cmselect/cmpubadm/334/33403.htm [accessed 5 May 2017].

Ipsos MORI. (2015). *How Britain voted in 2015: The 2015 Election – who voted for whom?* Ipsos MORI. Available from www.ipsos.com/ipsos-mori/en-uk/how-britain-voted-2015 [accessed 5 May 2017].

Jensen, C. and Spoon, J-J. (2011). Compelled without direction: compulsory voting and party system spreading. *Electoral Studies*, 30(4): 700–711.

King, M. (1968). 'The Other America'. Speech, Grosse Pointe High School, 14 March. Grosse Pointe Historical Society. Available from www.gphistorical.org/mlk/mlkspeech/mlk-gp-speech.pdf [accessed 16 November 2017].

Koch, I. (2016). Bread and butter politics: democratic disenchantment and everyday politics on an English council estate. *American Ethnologist*, 43(2): 282–294.

Lever, A. (2008). 'A liberal defence of compulsory voting': some reasons for scepticism. *Politics*, 28(1): 61–64.

Lyon, D. (2001). *Surveillance society: monitoring everyday life*. London: McGraw Hill.

Marsden, C. (2014). Hyper-power and private monopoly: The unholy marriage of (neo) corporatism and the imperial surveillance state. *Critical Studies in Media Communication*, 31(2): 100–108.

Newburn, T. (2015). The 2011 England riots in recent historical perspective. *British Journal of Criminology*, 55: 39–64.

Norris, P. (2003). 'Young people and political activism: from the politics of loyalties to the politics of choice?' Paper presented to the Council of Europe Symposium, *Young People and Democratic Institutions: From Disillusionment to Participation*, Strasbourg, 27–28 November 2003. Available from https://sites.hks.harvard.edu/fs/pnorris/Acrobat/COE.pdf [accessed 15 November 2017].

Parry, G., Moyser, G. and Day, N. (1992). *Political participation and democracy in Britain*. Cambridge: Cambridge University Press.

Phelps, E. (2012). Understanding electoral turnout among British young people: A review of the literature. *Parliamentary Affairs*, 65(1): 281–299.

Pickard, S. and Bessant, J. (eds). (2017). *Young people and re-generating politics in times of crises.* London: Palgrave Macmillan.

Putnam, R. (2004). *Bowling alone: The collapse and revival of American community.* New York: London: Simon & Schuster.

Scheuerman, W. (2016). Digital disobedience and the law. *New Political Science,* 38(3): 299–314.

Singh, S. (2014). Compulsory voting and the turnout decision calculus. *Political Studies,* 63(3): 548–568.

Staples, W. (1997). *The culture of surveillance: discipline and social control in the United States.* New York: St. Martin's Press.

Stoker, G. (2006). *Why politics matters: making democracy work.* Basingstoke: Palgrave Macmillan.

Stoycheff, E. (2016). Under surveillance: examining Facebook's spiral of silence effects in the wake of NSA internet monitoring. *Journalism & Mass Communication Quarterly,* 93(2): 296–311.

Tonge, J., Mycock, A. and Jeffery, B. (2012). Does citizenship education make young people better-engaged citizens? *Political Studies,* 60(3): 578–602.

Wagner, P. (2001). *A history and theory of the social sciences.* London: Sage.

Wagner, P. (2011). *The democratic crisis of capitalism: reflections on political and economic modernity in Europe.* Paper 44, LSE, London.

Wattenberg, M. (2008). *Is voting for young people?* New York: Pearson Longman.

Westholme, A. and Niemi, R. (1986). Youth unemployment and political alienation. *Youth and Society,* 18(1): 55–80.

Wilks-Heeg, S., Blick, A. and Crone, S. (2012). *How democratic is the UK? The 2012 audit.* Liverpool: Democratic Audit.

Younge, G. (2011). 'Reading the riots: investigating England's summer of disorder'. *Guardian* LSE, 21 September. Available from www.theguardian.com/commentis free/2011/aug/14/young-british-rioters-political-actions [accessed 16 November 2017].